Advances in Experimental Medicine and Biology

Volume 1014

More information about this series at http://www.springer.com/series/5584

Alireza Fazeli • William V. Holt
Editors

Periconception in Physiology and Medicine

Springer

Editors
Alireza Fazeli
Academic Unit of Reproductive and Developmental Medicine
University of Sheffield
Sheffield, UK

Institute of Biomedicine and Translational Medicine
Department of Pathophysiology, University of Tartu
Tartu, Estonia

William V. Holt
Academic Unit of Reproductive and Developmental Medicine
University of Sheffield
Sheffield, UK

ISSN 0065-2598 ISSN 2214-8019 (electronic)
Advances in Experimental Medicine and Biology
ISBN 978-3-319-62412-9 ISBN 978-3-319-62414-3 (eBook)
DOI 10.1007/978-3-319-62414-3

Library of Congress Control Number: 2017950292

Printed on acid-free paper

This Springer imprint is published by Springer Nature
The registered company is Springer International Publishing AG
The registered company address is: Gewerbestrasse 11, 6330 Cham, Switzerland

Preface

Reproductive science is undeniably important for the survival of all species on earth, but attempting to understand reproductive processes in animal species is a dauntingly complex task. The topic ranges from details of gametogenesis, to fertilisation and the subsequent processes of embryonic development, to growth and sexual differentiation, to endocrinology to aspects of behaviour and brain function. As if this list were not broad enough, evolution introduced many unexpected twists and turns as species colonised almost all potential habitats on earth and had to adapt to local conditions. Although such adaptations have been studied for many years, modern scientific advances have enabled us to drill down into the intricate details of gene expression, protein synthesis and the immune system, and it is increasingly clear that many unsuspected details of adaptation are now being uncovered. All animals evolve and adapt to their environment to optimise fitness, and the science of understanding these interactions has led to the realisation that phenomena, such as temperature, photoperiod and seasonality, have massive impacts on reproductive success. Reproductive success itself is not simply a matter of producing the maximum number of offspring, but ensuring that those offspring have the best chance for surviving and continuing to adapt in the future.

It is increasingly realised that parental diet, behaviour, exposure to stress and other influences experienced during the weeks and months leading up to, and including, the time of conception (known as the periconception period) can both help and hinder reproductive success. This is the key message presented in nearly every chapter of this book, and the motive for producing this book was to bring together a collection of authoritative articles that, while focused on different aspects of the periconception period, collectively emphasise the important influence of this period on health and the future well-being of offspring. To demonstrate that the periconception period is not solely a topic of interest in human medicine, we have included articles on a broad range of species in an effort to show that the topic is of universal relevance for reproductive science.

The editors are confident that this book is the first of a series of books that will try to answer some of the unanswered questions in the field and will further highlight the research needed to improve our knowledge regarding the periconception period and its associated events. We hope this book will be followed by other books dedicated to advances in periconception research and will further shed light on the events and mechanisms taking place during periconception period.

Aside from the editors, other important players in the conception and production of this book were the colleagues and friends gathered during the two successful EU-funded COST Actions, Gemini (FA0702; www.cost-gemini.eu) and Epiconcept (FA1201; www.cost-epiconcept.eu). Discussions held during the course of these two Actions, fostered close partnerships, collaborations and lasting friendships between colleagues of different nationalities and inspired new ideas forming the main structure of the current book. We would like to acknowledge the contributions and support of Ann Van Soom (Chair of Epiconcept) and Laszlo Tecsi (Administrator of Gemini and Epiconcept) who helped us during the production of this book. Your friendship is much appreciated. We are also grateful to Gill Burkinshaw (Departmental secretary) for her support and patience during the preparation of this book. Here we would also like to acknowledge the COST Organisation for supporting these actions as well as the help and support of Ioanna Stavridou (Science Officer of Gemini and Epiconcept) for her support and dedication to our actions.

Finally we would like to thank a great many colleagues that took part in the preparation of the manuscripts that form different chapters of this book and have helped the editors and authors with their critical and constructive comments.

Last but not least, we would like to thank our families and loved ones for sacrificing the quality time that we could have spent with them and allowing us to spend it on production of this book.

Sheffield, UK

Alireza Fazeli
William V. Holt

Contents

Contributors

Akwasi A. Amoako Department of Reproductive Medicine, Central Manchester, University Hospitals NHS Foundation Trust, St Mary's Hospital, Manchester, UK

Oleg Denisenko Department of Medicine, University of Washington, Seattle, WA, USA

Judith J. Eckert Faculty of Medicine, University of Southampton, Mailpoint 840, Southampton General Hospital, Southampton, UK

Alireza Fazeli Academic Unit of Reproductive and Developmental Medicine, University of Sheffield, Sheffield, UK

Institute of Biomedicine and Translational Medicine, Department of Pathophysiology, University of Tartu, Tartu, Estonia

Bernd Fischer Department of Anatomy and Cell Biology, Martin Luther University Faculty of Medicine, Halle, Germany

Tom P. Fleming Biological Sciences, University of Southampton, Mailpoint 840, Southampton General Hospital, Southampton, UK

Jacqueline Gürke Department of Anatomy and Cell Biology, Martin Luther University Faculty of Medicine, Halle, Germany

Katarzyna Joanna Grybel Department of Anatomy and Cell Biology, Martin Luther University Faculty of Medicine, Halle, Germany

Elisa Haucke Department of Anatomy and Cell Biology, Martin Luther University Faculty of Medicine, Halle, Germany

Sonia Heras Department of Reproduction, Obstetrics and Herd Health, Faculty of Veterinary Medicine, Ghent University, Merelbeke, Belgium

William V. Holt Academic Unit of Reproductive and Developmental Medicine, University of Sheffield, Sheffield, UK

J.L.M.R. Leroy University of Antwerp, Campus Drie Eiken, Wilrijk, Belgium

Patrick Lonergan School of Agriculture and Food Science, University College Dublin, Belfield, Ireland

Emma S. Lucas Division of Reproductive Health, Clinical Science Research Laboratories, Warwick Medical School, University of Warwick, Coventry, UK

Veronica Maillo Departamento de Reproducción Animal, Instituto Nacional de Investigación y Tecnología Agraria y Alimentaria (INIA), Madrid, Spain

S. Mareike Pendzialek Department of Anatomy and Cell Biology, Martin Luther University Faculty of Medicine, Halle, Germany

Tamer M. Nafee Sheffield Teaching Hospitals NHS Foundation Trust, University of Sheffield, Sheffield, UK

Bolarinde Ola Sheffield Teaching Hospitals NHS Foundation Trust, University of Sheffield, Sheffield, UK

Geert Opsomer Department of Reproduction, Obstetrics and Herd Health, Faculty of Veterinary Medicine, Ghent University, Merelbeke, Belgium

James Ord Department of Animal and Plant Sciences, University of Sheffield, Western Bank, Sheffield, UK

Academic Unit of Reproductive and Developmental Medicine, University of Sheffield, Sheffield, UK

Dimitrios Rizos Departamento de Reproducción Animal, Instituto Nacional de Investigación y Tecnología Agraria y Alimentaria (INIA), Madrid, Spain

Maria-Jesús Sánchez-Calabuig Departamento de Reproducción Animal, Instituto Nacional de Investigación y Tecnología Agraria y Alimentaria (INIA), Madrid, Spain

Anne Navarrete Santos Department of Anatomy and Cell Biology, Martin Luther University Faculty of Medicine, Halle, Germany

Maria Schindler Department of Anatomy and Cell Biology, Martin Luther University Faculty of Medicine, Halle, Germany

René Thieme Department of Anatomy and Cell Biology, Martin Luther University Faculty of Medicine, Halle, Germany

Mieke Van Eetvelde Department of Reproduction, Obstetrics and Herd Health, Faculty of Veterinary Medicine, Ghent University, Merelbeke, Belgium

Ann Van Soom Department of Reproduction, Obstetrics and Herd Health, Faculty of Veterinary Medicine, Ghent University, Merelbeke, Belgium

Adam J. Watkins Aston Research Centre for Healthy Ageing, School of Life and Health Sciences, Aston University, Birmingham, UK

Penelope J. Watt Department of Animal and Plant Sciences, University of Sheffield, Western Bank, Sheffield, UK

Introduction: A Brief Guide to the Periconception Environment

Alireza Fazeli and William V. Holt

Abstract Definition of the periconception period is not an exact science and is probably somewhat arbitrary. One can define it as spanning the period from the final stages of gamete maturation until formation of the embryo and the stages of embryonic development and implantation. Hence, the periconception period includes periods when spermatozoa are in the female reproductive tract, oocytes are matured and ovulated into the oviduct, fertilization occurs and the embryo undergoes development. By definition the implantation process and the early stages of placenta formation are also regarded as a part of the periconception period. In this article we highlight a few of the major advances which have transformed this topic over the last two decades. It is now clear that the fitness and wellbeing of developing mammalian embryos, including the human, are highly dependent on the health status, diet and habits of both parents especially in the months and weeks that precede the formation of oocytes and spermatozoa.

Keywords Development • Sex determination • Environment • Inheritance • Aquatic organisms • Fishes • Reptiles

---

A. Fazeli
Academic Unit of Reproductive and Developmental Medicine, University of Sheffield, Level 4, Jessop Wing, Tree Root Walk, Sheffield S10 2SF, UK

Institute of Biomedicine and Translational Medicine, Department of Pathophysiology, University of Tartu, 14B Ravila, Tartu 50411, Estonia
e-mail: A.Fazeli@sheffield.ac.uk; Alireza.Fazeli@ut.ee

W.V. Holt (✉)
Academic Unit of Reproductive and Developmental Medicine, University of Sheffield, Level 4, Jessop Wing, Tree Root Walk, Sheffield S10 2SF, UK
e-mail: Bill.Holt@sheffield.ac.uk

A. Fazeli, W.V. Holt (eds.), *Periconception in Physiology and Medicine*,
Advances in Experimental Medicine and Biology 1014,
DOI 10.1007/978-3-319-62414-3_1

1 Introduction

Our current understanding of the events that take place during the periconception period, such as sperm capacitation and early embryonic developmental processes, were gained during the course of the twentieth century. The majority of research on the periconception environment during the second half of the twentieth century focused on understanding the main constituents of the periconception environment, and how this environment could be replicated in vitro in order to support gamete maturation, in vitro fertilization (IVF) and embryonic development. The birth of Louise Brown (the first baby born by IVF) was a major indication of achieving this goal and signalled the ability of human kind to replicate the periconception environment in vitro to a level sufficient to support the final stages of gamete maturation, conception and early embryonic development to such an extent that successful conception and the production of a conceptus capable of further development to term in vitro was achieved.

Without doubt, the success of IVF was a landmark achievement in reproductive research history and helped with the conception of babies for many infertile couples. However, it created a belief among reproductive physiologists/experts that the periconception environment could be replaced by a relatively simple, buffered balanced salt solution media containing an adequate energy source. As IVF became more and more successful and widespread, inevitably its success resonated the belief that, just as IVF media is a simple and static environment, the periconception milieu in vivo should also be a passive milieu. Once gametes enter this environment, they should doubtless initiate the final stages of maturation that lead to fertilization and early embryonic development. It became generally accepted and constituted a common dogma that the contributions of reproductive physiology towards the preconception milieu must be relatively negligible and limited in providing an energy source, the right pH, temperature and osmotic pressure. This view characterized the oviduct/Fallopian tubes and upper parts of the female reproductive tract as passive conduits, and the processes of fertilization and embryonic development as autonomous events with minimal need of cues or signals conveyed from their surrounding environment. This view supposed that once these processes have been initiated they can progress on their own.

In this chapter, we will present a summary of the work that challenged this view and established our current knowledge and modern physiological understanding of the periconception environment. We will first address the historic perspectives and experiments challenging the view that the periconception environment is a static environment. Thereafter we will explain some seminal experiments describing the major elements involved in forming the periconception milieu. We will then continue to bring examples of how a dynamic periconception milieu can either influence function and physiology of gametes as well as the development of embryos. Finally we will focus our attention on different possibilities to understand how the periconception milieu is regulated.

The majority of current advances in understanding the periconception environment resulted from work performed on defining the composition of embryo culture media for optimum embryonic development. This information is reviewed in the manuscript by Dimitrios Rizos et al. that can be found elsewhere in this volume.

1.1 Discovery of a Dynamic Periconception Milieu

After sexual intercourse, the ejaculated spermatozoa must focus their efforts upon reaching the uterus, oviducts and ultimately the newly ovulated eggs. Their uniquely adapted cellular structure gives them the ability to propel themselves forwards through the viscous milieu of the female reproductive tract, but their journey towards the egg is not as straightforward as driving a car from one place to another, where the car itself is not changed by its environment and in turn it does not change the road surface or the prevailing weather conditions. The spermatozoa have an entirely different prospect before them when they begin their journey. They must survive the rigorous selection processes imposed by the female reproductive tract (Holt and Fazeli 2016) before they reach the vicinity of the egg, and most of them fail their task (Holt 2009). There are many reasons for this: different regions of the female reproductive tract, i.e. the vagina, cervix, uterus and oviducts, create their own local environments. These are known to differ in simple characteristics such as pH or osmolarity, but also in more complex attributes such as ionic composition, presence or absence of hormones and the local identity of proteins in the luminal fluids. Spermatozoa are known to respond sensitively to changes in their environment by changing their motion pattern, sometimes becoming quiescent (Overstreet and Cooper 1975; Suarez et al. 1992), more active (Wennemuth et al. 2003) or changing the shape of their swimming trajectories (Armon and Eisenbach 2011; McNutt et al. 1994; Suarez and Osman 1987). Besides changing their motion characteristics, the spermatozoa also respond to external stimuli by adjusting their ability to undergo membrane fusion. Achieving an appropriate degree of sperm plasma membrane fluidity is crucially important both for the prevention of inappropriate and premature acrosome reactions and allowing the acrosome reaction to occur when a spermatozoon is poised and ready to commence the fertilization process.

Since Overstreet and Cooper's (Overstreet and Cooper 1975) original observations concerning sperm motility suppression by the oviduct, there has been considerable interest in many aspects of sperm-oviduct interactions especially the formation of a sperm reservoir within the oviduct, where a highly selected cohort of spermatozoa is able to reside for a few hours or days prior to the events of fertilization. Many in vitro experiments had shown that if spermatozoa were co-incubated with oviductal epithelial cells, even if they were from a different species (Ashizawa and Nishiyama 1983a, b), their survival was enhanced considerably. It was also observed that the spermatozoa usually became firmly bound to the epithelial cell surface, demonstrating the importance of membrane-membrane contact in the maintenance of sperm viability within the oviductal sperm reservoir (Dobrinski et al. 1997; Smith and Nothnick 1997; Boilard et al. 2001; Suarez and Pacey 2006).

The membrane-membrane contact was shown to be involved in prolonging the duration of sperm survival and initiated protein synthesis (Ellington et al. 1993; Thomas et al. 1995). Collectively these studies have revealed that these interactions are more complex than hitherto suspected and that cell signalling mechanisms play a major part in the sperm-oviduct dialogue.

That spermatozoa are able to elicit de novo gene expression in oviductal cells was first noticed in experiments carried out by Ellington and her colleagues (Ellington et al. 1993) using bovine tissues. The sophisticated proteomic methods available today had not yet been developed and therefore the observations were based on the appearance of new protein spots in electrophoresis gels, but only if the oviductal cells had been co-incubated with spermatozoa. At the time, these results were somewhat controversial and some researchers attributed them to various technical artefacts. However, later research vindicated the authors and confirmed their observations in a variety of ways. Fazeli and colleagues (Fazeli et al. 2004) used microarrays to show that novel gene transcription occurred in the mouse uterus/oviductal complex after natural mating. Using as controls a line of mutant male mice (T145H) that could not produce spermatozoa, but could mate naturally and produce seminal plasma, confirmed that the novel gene expression was a response to the spermatozoa. Although it was not possible at the time to identify all of the novel transcripts, two were identified as adrenomedullin and prostaglandin endoperoxidase synthase 2. It is of interest that adrenomedullin has since been implicated in the modulation of ciliary movement both in epithelia, such as the oviduct and nasal passage, as well as the sperm flagellum itself (Liao et al. 2011, 2012; Li et al. 2010; Chiu et al. 2010).

These studies were followed by experiments showing that the production of 19 novel proteins was identifiable if spermatozoa alone were introduced into isolated porcine oviducts and then incubated at body temperature for 2 h (Georgiou et al. 2005). A highly significant aspect of the data from this study was the identification of heat shock protein 70 (Hsp70) among the soluble components of the luminal fluid. Until relatively recently it was believed that mammalian HSPs are exclusively intracellular molecules and that they are only present in extracellular compartments in pathological conditions such as necrotic cell death. However, there is now extensive evidence to support the view that stress proteins can be released under non-pathological conditions and exert protective roles. The presence of heat shock proteins within oviductal fluid is now well established and they are known to exert influence on sperm-oocyte binding in humans (Marin-Briggiler et al. 2010).

Once these pathfinding experiments had been carried out, the methodology for investigating sperm-oviduct interactions could be characterized (Aldarmahi et al. 2012; Yeste et al. 2014) in order to answer specific questions. It is now evident that if direct contact between spermatozoa and epithelia is prevented, the induction of novel gene expression does not take place (Yeste et al. 2009a), and this correlates with the observation that the improved sperm survival seen in coculture no longer happens (Yeste et al. 2009b).

2 How Does a Dynamic Periconception Milieu Affect Events During Periconception?

The significance of sperm-induced gene expression in the oviduct is still poorly understood with respect to its effects upon the periconception environment and the future fitness of the developing embryo. A recent review of oviductal function (Li and Winuthayanon 2017) emphasized that the oviductal environment has a crucially important role in providing embryos with appropriate metabolic support and protecting the embryo from oxidative and other stresses. As fertilization itself can take place in vitro, it is clear that the presence of these novel proteins is not obligatory for this process; however, they may help to prepare the oviduct for the future arrival and development of the newly formed embryo.

Studies of the proteins found in the oviductal fluids of domestic animals have produced a long list of candidates (see, e.g. Coy and Aviles 2010; Coy et al. 2008a, b; Das et al. 2013; Seytanoglu et al. 2008). Some of them have been functionally characterized and are known to affect processes such as the acrosome reaction, sperm motility and sperm-oocyte binding. Exposure to oviductal fluid of newly ovulated oocytes causes hardening of the zona pellucida and inhibition of polyspermy (Coy and Aviles 2010). One protein worthy of special mention is oviductin, or oviduct-specific glycoprotein (OVGP1) (McCauley et al. 2003; Yang et al. 2015; Zhao et al. 2016). This major component of oviductal fluid is a high-molecular-weight, oviduct-specific and oestrogen-dependent glycoprotein that has been identified in a variety of species, including the mouse, hamster, rabbit, cow, pig, baboon, rhesus monkey, goat and human. It is one of the set of proteins that are upregulated by the arrival of spermatozoa, and logically therefore, it may be hypothesized that it may have a role over and above the modulation of sperm function. This proposal is supported by studies in which embryo cultures have been supplemented with recombinant oviductin (Coy and Yanagimachi 2015) and shown to enhance aspects of embryonic development. For example, in a study of feline embryonic development, the relative mRNA abundance of GJA1, a gene, whose expression level is known to be a marker of embryo quality, was significantly increased in blastocysts after oviductin treatment (Hribal et al. 2014). In contrast to this, expression of OCT4, HSP70, DNMT1, DNMT3A, BAX, IGF1R and GAPDH was not significantly affected.

These studies provide useful clues to the roles and importance of oviductal proteins in relation to embryo quality and development, but in order to understand their exact physiological function, we need to know the details, such as the amount of protein released to the periconception milieu and how much of these proteins come into contact with gametes and embryos.

3 Spatial Formation of Oviduct and Uterine Cavity Affecting the Periconception Milieu and Its Dynamic Nature

In nearly all mammals the periconception milieu concerns mostly the upper parts of the female reproductive tract. In the past some investigators have investigated the nature of secretions produced by the female tract at different stages of the reproductive cycle (Grippo et al. 1992; Killian 2011; Leese et al. 2008; Hunter 2012), and some investigators have examined the way in which the cellular proteins change during the oestrous cycle (Sostaric et al. 2005; Seytanoglu et al. 2008). Few precise measurements have been carried out to determine how protein production and other potential elements produced by the oviduct and uterine cavity change in response to gametes and embryos. Two recent studies are worthy of note, however. Changes in pig oviductal gene transcription in vivo following artificial insemination (López-Úbeda et al. 2015) found that 17 genes were upregulated following insemination and 9 were downregulated. Analysis of the upregulated genes, carried out by examining them in the context of functional networks revealed that the pathways affected by insemination were related to the inflammatory response, the immune system, to molecular transport, protein trafficking and developmental disorder and to cell-to-cell signalling and interactions. The authors concluded that the changes probably represent a degree of preparation for the imminent events of fertilization and early embryo development. At this stage it is difficult to go further than this and focus on more detail. However, a complementary in vivo study (Almiñana et al. 2014) showed if pig oviducts were surgically inseminated with sex-sorted spermatozoa, where the separate populations of X- and Y-bearing spermatozoa were inseminated into different oviducts in the same animals, the gene upregulation responses were different. The significance of the latter study is difficult to interpret, although it shows clearly that the oviductal epithelium has a sophisticated ability to scrutinize the cell surfaces of the inseminated spermatozoa and react in different ways.

4 Periconception Milieu and DOHAD

The relationships between the intrauterine environment and the embryo during the early life of mammals and the onset of adult diseases, such as cardiovascular disease, hypertension and diabetes were initially identified from epidemiological observations on human populations, and have since been confirmed in experimental mammals (Barker et al. 2010; Thornburg et al. 2010). These observations, often collectively known as the Barker hypothesis, repeatedly show that if embryos undergo different forms of 'stress' during early development, children are likely to be underweight at birth and will then show phenotypic symptoms of disease as they develop into adults. Many of the articles presented in this book underline both the mechanisms and consequences that underpin the hypothesis that maternal nutritional status has long-lasting impacts on the future health and wellbeing of her offspring. Most attention has been paid to dietary stress, where the embryo initially seems to adapt its metabolic functions to make the most of the limited resources available. If conditions improve later in life, the individual cannot cope with the better lifestyle and tends to become obese and develop a suite of late-onset diseases.

However, it is also worth asking whether a suboptimal periconception environment might have negative consequences for embryonic development from as early in development as spermatogenesis (i.e. the paternal periconception environment) and the events surrounding fertilization. A recent study conducted within the context of a programme aimed at investigating possible links between diet and human fertility (Faure et al. 2015) compared semen parameters from 92 subfertile men with 91 fertile men and found that the extent of sperm DNA fragmentation was positively associated with birth weight and inversely correlated with total sperm count. As birth weight in the fertile population was significantly lower than in infertile patients, the authors suggested that paternal characteristics may even have affected visceral fat deposits in the newborns. Paternal body weight itself has been implicated as a determinant of deleterious epigenetic effects. Soubry and colleagues (Soubry 2015; Soubry et al. 2014, 2016) found significant differences in DNA methylation at differentially methylated regions (DMRs) of several imprinted genes if the father was obese.

Logically the presence of endocrine disrupting chemicals (EDCs) within the foetal environment, embryo, newborn or juveniles and the female reproductive tract could exert additional stress or have important influences on embryonic growth and development. This topic was recently reviewed in detail (Ho et al. 2017). A range of toxic persistent organic pollutants (POPs), such as polychlorinated biphenyls (PCBs), polychlorinated dibenzop-dioxins (PCDDs), polychlorinated dibenzofurans (PCDFs), dichlorodiphenyldichloroethylene (p,p'-DDE) and polybrominated diphenyl ethers (PBDEs), continue to contaminate the environment due to past and present human activities. Recent evidence from a Swedish human population showed that prenatal exposure to polychlorinated biphenyls (PCBs) was associated with higher birth weight, and PBDE exposure with lowered birth weight (Lignell et al. 2013). Although these effects are complex and difficult to clarify, such human population studies suggest that wild species, whose body burden of such chemicals is often higher, will also be affected. Extensive and detailed data on contaminant concentrations in arctic wild species, including mammals, birds and fishes, reported by (Letcher et al. 2010) lend support to this idea. In fact, these authors commented that evidence of defective neurological development in some polar bears might be attributed to such long-term effects, but they could not be sure because of the difficulties involved in obtaining relevant data.

Although still sparse, evidence from wild species should not be overlooked when considering the likely impacts of pollutants on human reproduction. Chemical analyses of several large mammals (seals, porpoises, whales and polar bears) have demonstrated high body burdens of hydrophobic contaminants, such as PCBs and brominated flame retardants (for review, Sonne 2010; Sonne et al. 2006). Typically, these species acquire pollutants via their diet and bioaccumulate EDCs, which tend to be lipophilic compounds, within their body fat so that the concentrations increase together with increasing age. When females begin to suckle their offspring, their milk is enriched with EDCs, and the EDCs are transferred to their newborns, with probable effects on survival and reproductive development. Surprisingly, however, it has proven difficult to demonstrate that PCBs and other lipophilic compounds actually cause impaired reproductive development (Letcher et al. 2010), although population and pharmacokinetic modelling studies of East Greenland polar bears, backed up by experimental studies of Greenland sledge dogs, predict the imminent occurrence of negative population impacts (Sonne 2010). Evidence for reduced female reproductive efficacy has also been found in studies of the European harbour porpoise and short-beaked common dolphin (Munson et al. 1998). In harbour porpoises, high persistent organic pollutant burdens tended to be associated with few ovarian scars, suggesting that high contaminant levels may be inhibiting ovulation; however, the significance of ovarian scars has recently been re-evaluated (Pierce et al. 2008).

5 Periconception and Epigenetic Mechanisms

Epigenetic markers are enzyme-mediated chemical modifications of DNA and of its associated chromatin proteins. These modifications play key roles in regulating genomic functions, without changing the primary DNA sequence, and are then passed down through many cell generations. Epigenetic marks influence the way in which chromatin is transcribed and are therefore key determinants of embryogenesis and development. The main epigenetic mechanisms involve DNA methylation, histone modifications and microRNA species, which are involved in genomic activity states.

5.1 DNA Methylation

DNA methylation is the most extensively-studied epigenetic modification and changing DNA methylation patterns in key elements of a gene, such as promoters and enhancers, can have a profound effect on gene function. The best characterized examples of methylation-mediated silencing mechanisms are genomic imprinting, X-chromosome inactivation and silencing of retrotransposons. Imprinted genes specifically have one or more transcripts that are expressed preferentially or exclusively from one parental allele. This monoallelic expression is likely to be initiated by differential methylation of the oocytes and spermatozoa. Shortly after fertilization (the periconception period) this differential methylation leads to additional molecular changes, including allele-specific methylation, which ultimately results in gene expression exclusively from one allele.

Maintenance of DNA methylation is mediated by DNA methyltransferases (DNMTs), and this process is relatively well understood at a biochemical level. In contrast, much remains to be learned about the establishment of DNA methylation (de novo methylation) and the mechanisms underlying its deregulation.

As these mechanisms operate in strictly-defined stages of development, it is possible that the periconception period is split into several particularly sensitive windows of vulnerability. In the context of maternal nutrition and early development, it is notable that DNA methylation is dependent on the one-carbon metabolism pathway. This system relies on a variety of enzymes, whose activities, in turn, depend on micronutrients supplied through the diet. For example, methionine is used by DNA methyltransferases to attach methyl groups to the carbon-5 position of cytosine bases, thus generating 5-methyl cytosine. Different nutrients involved in one-carbon metabolism, such as folate, vitamin B6, vitamin B12, choline and methionine, play an important role in DNA methylation through their influence on S-adenosylmethionine and the methyltransferase inhibitor S-adenosylhomocysteine.

5.2 microRNAs

The human genome contains only 20,000 protein-coding genes, representing <2% of the total genome, whereas a substantial fraction of the human genome can be transcribed, yielding many short- or noncoding-RNAs (ncRNAs) with limited protein-coding capacity. Among these, the most extensively-studied ncRNAs are microRNAs (miRNAs), which are evolutionarily conserved and located within the introns and exons of protein-coding genes or in intergenic regions. MicroRNA species have been detected in spermatozoa where their functions are only now being elucidated (Miller et al. 2010; Miller and Ostermeier 2006); it is, however, interesting to note that differences between fertile and infertile men have been detected in the complement of sperm miRNA (Miller et al. 2005; Ostermeier et al. 2002, 2004). Dietary factors have been shown to modify miRNA expression profiling, notably those associated with maternal nutrition, insulin resistance and inflammation (for review, see Lee 2015).

Small RNAs are believed to direct DNA methylation machinery or chromatin/histone modification complexes for targeting specific genomic loci. The unique sequences of ncRNAs ensure the site specificity of de novo epigenetic silencing (Stephane and Robert 2013). The sperm-borne miRNA profiles are sensitive to various environmental exposures/conditions including paternal obesity in mice, chronic stress in male mice and smoking in humans (Marczylo et al. 2012), and given that miRNA enters the oocyte during fertilization (Rivera and Ross 2013; Amanai et al. 2006), it would be surprising if they did not exert any impacts upon the next generation. Changes in sperm miRNA content are indeed correlated with transgenerational behavioural effects in experimental animals (Guerrero-Bosagna et al. 2012; Anway et al. 2006) and metabolic outcomes (Vrijens et al. 2015). Exposure to the fungicide vinclozolin has been shown to induce transgenerational changes in specific miRNAs in primordial germ cells in F1 to F3 generations.

Like the miRNA of spermatozoa, it is becoming clear that miRNAs also play important roles in oocytes and developing embryos (Zhao and Srivastava 2007). Optimum regulation of genes or critical gene regulatory events associated with early embryonic development have been shown to be controlled by miRNAs. For example, miR-430 has been shown to promote deadenylation and clearance of maternal mRNAs in zebrafish oocytes (Mishima et al. 2006; Schier and Giraldez 2006).

This brief summary has illustrated how the field of miRNA biology has been transformed over the past 25 years or thereabouts, from a topic that was largely mysterious, or even treated with scepticism, into a massively important subject that has important implications for many aspects of developmental biology and also for the treatment or prevention of disease. There may also be opportunities to modify miRNAs in the context of assisted reproductive technologies, in an effort to correct some aspects of inherited disease (Martin et al. 2017; Bouckenheimer et al. 2016).

6 Conclusions

The impact of the periconception environment on the future health of newborns was not widely appreciated until about 25 years ago, but nowadays it is increasingly recognized. The outcomes can be both positive and negative for the individuals involved. As income and living standards in some populations around the world have increased, it is apparent that they are accompanied by a significantly increased incidence of diabetes; in fact, India has earned the nickname 'diabetes capital of the world' (Wells et al., 2016). A likely reason for this outcome is an inability to cope with high glucose intakes in adulthood, which fail to match the way that the population is historically and biologically adapted to survive on a poorer quality diet. On the other hand, the studies on pollutant exposure and smoking discussed above, and possibly also some of the pharmaceuticals used in human medicine, indicate strongly that the reproductive system is a sensitive target likely to be affected by changing environments in ways that might be unexpected and unpredictable. These observations pose ethical questions for governments, especially in terms of agriculture and public health, where pesticide use is widely regarded as an essential way to improve crop yields and prevent the spread of malaria and other insect-borne diseases. It is likely that the cure also causes problems further down the line! Answering these questions is outside the scope of this book, but the chapters we present should at least help to raise relevant questions.

References

Aldarmahi A, Elliott S, Russell J, Klonisch T, Hombach-Klonisch S, Fazeli A (2012) Characterisation of an *in vitro* system to study maternal communication with spermatozoa. Reprod Fertil Dev 24(7):988–998. http://dx.doi.org/10.1071/RD11268

Almiñana C, Caballero I, Heath PR, Maleki-Dizaji S, Parrilla I, Cuello C, Gil MA, Vazquez JL, Vazquez JM, Roca J, Martinez EA, Holt WV, Fazeli A (2014) The battle of the sexes starts in the oviduct: modulation of oviductal transcriptome by X and Y-bearing spermatozoa. BMC Genomics 15:293. doi:10.1186/1471-2164-15-293

Amanai M, Brahmajosyula M, Perry N (2006) A restricted role for sperm-borne microRNAs in mammalian fertilization. Biol Reprod 75(6):877–884. doi:10.1095/biolreprod.106.056499

Anway MD, Memon MA, Uzumcu M, Skinner MK (2006) Transgenerational effect of the endocrine disruptor vinclozolin on male spermatogenesis. J Androl 27(6):868–879

Armon L, Eisenbach M (2011) Behavioral mechanism during human sperm chemotaxis: involvement of hyperactivation. PLoS One 6(12):e28359. doi:10.1371/journal.pone.0028359

Ashizawa K, Nishiyama H (1983a) Effects of oviducal cells on the maintenance of motility and fertilizing capacity of fowl spermatozoa stored in a diffusion chamber. Poult Sci 62(11):2276–2279

Ashizawa K, Nishiyama H (1983b) Prolonged survival of fowl spermatozoa in the oviducal tissues in organ culture. Br Poult Sci 24(1):27–32. doi:10.1080/00071668308416710

Barker DJ, Gelow J, Thornburg K, Osmond C, Kajantie E, Eriksson JG (2010) The early origins of chronic heart failure: impaired placental growth and initiation of insulin resistance in childhood. Eur J Heart Fail 12(8):819–825. doi:10.1093/eurjhf/hfq069. hfq069 [pii].

Boilard M, Reyes-Moreno C, Sirard MA (2001) Binding of chaperonins to bovine spermatozoa by direct contact to apical plasma membrane of oviduct epithelial cells. Biol Reprod 64:112–112

Bouckenheimer J, Assou S, Riquier S, Hou C, Philippe N, Sansac C, Lavabre-Bertrand T, Commes T, Lemaitre JM, Boureux A, De Vos J (2016) Long non-coding RNAs in human early embryonic development and their potential in ART. Hum Reprod Update 23(1):19–40. doi:10.1093/humupd/dmw035

Chiu PC, Liao S, Lam KK, Tang F, Ho JC, Ho PC, WS O, Yao YQ, Yeung WS (2010) Adrenomedullin regulates sperm motility and oviductal ciliary beat via cyclic adenosine 5′-monophosphate/protein kinase A and nitric oxide. Endocrinology 151(7):3336–3347. doi:10.1210/en.2010-0077

Coy P, Avilés M (2010) What controls polyspermy in mammals, the oviduct or the oocyte? Biol Rev 85:593–605

Coy P, Canovas S, Mondejar I, Saavedra MD, Romar R, Grullon L, Matas C, Aviles M (2008a) Oviduct-specific glycoprotein and heparin modulate sperm-zona pellucida interaction during fertilization and contribute to the control of polyspermy. Proc Natl Acad Sci U S A 105(41):15809–15814

Coy P, Lloyd R, Romar R, Matas C, Adan AG, Holt WV (2008b) Bovine oviductal fluid affects quality and gene expression of in vitro derived porcine blastocysts. Reprod Domest Anim 43:66–66

Coy P, Yanagimachi R (2015) The common and species-specific roles of oviductal proteins in mammalian fertilization and embryo development. Bioscience 65(10):973–984. doi:10.1093/biosci/biv119

Das SK, Sharma AK, Mohapatra SK, Bhatia V, Chatterjee A, Mohanty AK (2013) Purification of cattle oviduct specific proteins and their effect on in vitro embryo development. Livest Sci 152(1):88–93. https://doi.org/10.1016/j.livsci.2012.12.002

Dobrinski I, Smith TT, Suarez SS, Ball BA (1997) Membrane contact with oviductal epithelium modulates the intracellular calcium concentration of equine spermatozoa in vitro. Biol Reprod 56(4):861–869

Ellington JE, Ignotz GG, Ball BA, Meyers-Wallen VN, Currie WB (1993) De novo protein synthesis by bovine uterine tube (oviduct) epithelial cells changes during co-culture with bull spermatozoa. Biol Reprod 48(4):851–856

Faure C, Dupont C, Chavatte-Palmer P, Gautier B, Levy R (2015) Are semen parameters related to birth weight? Fertil Steril 103(1):6–10. https://doi.org/10.1016/j.fertnstert.2014.11.027
Fazeli A, Affara NA, Hubank M, Holt WV (2004) Sperm-induced modification of the oviductal gene expression profile after natural insemination in mice. Biol Reprod 71(1):60–65. doi:10.1095/biolreprod.103.026815
Georgiou AS, Sostaric E, Wong CH, Snijders AP, Wright PC, Moore HD, Fazeli A (2005) Gametes alter the oviductal secretory proteome. Mol & cell proteomics : MCP 4(11):1785–1796. doi:10.1074/mcp.M500119-MCP200
Grippo AA, Henault MA, Anderson SH, Killian GJ (1992) Cation concentrations in fluid from the oviduct ampulla and isthmus of cows during the estrous-cycle. J Dairy Sci 75:58–65
Guerrero-Bosagna C, Covert TR, Haque MM, Settles M, Nilsson EE, Anway MD (2012) Epigenetic transgenerational inheritance of vinclozolin induced mouse adult onset disease and associated sperm epigenome biomarkers. Reprod Toxicol 34(4):694–707
Ho S-M, Cheong A, Adgent MA, Veevers J, Suen AA, Tam NNC, Leung Y-K, Jefferson WN, Williams CJ (2017) Environmental factors, epigenetics, and developmental origin of reproductive disorders. Reprod Toxicol 68:85–104. https://doi.org/10.1016/j.reprotox.2016.07.011
Holt WV (2009) Is semen analysis useful to predict the odds that the sperm will meet the egg? Reprod Domest Anim 44(Supplement 3):31–38
Holt WV, Fazeli A (2016) Sperm selection in the female mammalian reproductive tract. Focus on the oviduct: hypotheses, mechanisms, and new opportunities. Theriogenology 85(1):105–112. doi:10.1016/j.theriogenology.2015.07.019
Hribal R, Hachen A, Jewgenow K, Zahmel J, Fernandez-Gonzalez L, Braun BC (2014) The influence of recombinant feline oviductin on different aspects of domestic cat (Felis catus) IVF and embryo quality. Theriogenology 82(5):742–749. https://doi.org/10.1016/j.theriogenology.2014.06.009
Hunter RH (2012) Components of oviduct physiology in eutherian mammals. Biol Rev Camb Philos Soc 87(1):244–255. doi:10.1111/j.1469-185X.2011.00196.x
Killian G (2011) Physiology and endocrinology symposium: evidence that oviduct secretions influence sperm function: a retrospective view for livestock. J Anim Sci 89(5):1315–1322. doi:10.2527/jas.2010-3349
Lee H-S (2015) Impact of maternal diet on the epigenome during in utero life and the developmental programming of diseases in childhood and adulthood. Forum Nutr 7(11):9492–9507. doi:10.3390/nu7115467
Leese HJ, Hugentobler SA, Gray SM, Morris DG, Sturmey RG, Whitear SL, Sreenan JM (2008) Female reproductive tract fluids: composition, mechanism of formation and potential role in the developmental origins of health and disease. Reprod Fertil Dev 20(1):1–8
Letcher RJ, Bustnes JO, Dietz R, Jenssen BM, Jorgensen EH, Sonne C, Verreault J, Vijayan MM, Gabrielsen GW (2010) Exposure and effects assessment of persistent organohalogen contaminants in arctic wildlife and fish. Sci Total Environ 408(15):2995–3043. doi:10.1016/j.scitotenv.2009.10.038
Li HW, Liao SB, Chiu PC, Tam WW, Ho JC, Ng EH, Ho PC, Yeung WS, Tang F, WS O (2010) Expression of adrenomedullin in human oviduct, its regulation by the hormonal cycle and contact with spermatozoa, and its effect on ciliary beat frequency of the oviductal epithelium. J Clin Endocrinol Metab 95(9):E18–E25. doi:10.1210/jc.2010-0273
Li S, Winuthayanon W (2017) Oviduct: roles in fertilization and early embryo development. J Endocrinol 232(1):R1–R26. doi:10.1530/JOE-16-0302
Liao SB, Ho JC, Tang F, O WS (2011) Adrenomedullin increases ciliary beat frequency and decreases muscular contraction in the rat oviduct. Reproduction 141(3):367–372. doi:10.1530/REP-10-0230
Liao SB, Li HW, Ho JC, Yeung WS, Ng EH, Cheung AN, Tang F, WS O (2012) Possible role of adrenomedullin in the pathogenesis of tubal ectopic pregnancy. J Clin Endocrinol Metab 97(6):2105–2112. doi:10.1210/jc.2011-3290
Lignell S, Aune M, Darnerud PO, Hanberg A, Larsson SC, Glynn A (2013) Prenatal exposure to polychlorinated biphenyls (PCBs) and polybrominated diphenyl ethers (PBDEs) may influence

birth weight among infants in a Swedish cohort with background exposure: a cross-sectional study. Environ Health 12:44. doi:10.1186/1476-069x-12-44

López-Úbeda R, García-Vázquez FA, Romar R, Gadea J, Muñoz M, Hunter RHF, Coy P (2015) Oviductal transcriptome is modified after insemination during spontaneous ovulation in the sow. PLoS One 10(6):e0130128. doi:10.1371/journal.pone.0130128

Marczylo EL, Amoako AA, Konje JC, Gant TW, Marczylo TH (2012) Smoking induces differential miRNA expression in human spermatozoa: a potential transgenerational epigenetic concern? Epigenetics 7(5):432. doi:10.4161/epi.19794

Marin-Briggiler CI, Gonzalez-Echeverria MF, Munuce MJ, Ghersevich S, Caille AM, Hellman U, Corrigall VM, Vazquez-Levin MH (2010) Glucose-regulated protein 78 (Grp78/BiP) is secreted by human oviduct epithelial cells and the recombinant protein modulates sperm-zona pellucida binding. Fertil Steril 93(5):1574–1584. doi:10.1016/j.fertnstert.2008.12.132

Martin JH, Bromfield EG, Aitken RJ, Nixon B (2017) Biochemical alterations in the oocyte in support of early embryonic development. Cell Mol Life Sci 74(3):469–485. doi:10.1007/s00018-016-2356-1

McCauley TC, Buhi WC, GM W, Mao J, Caamano JN, Didion BA, Day BN (2003) Oviduct-specific glycoprotein modulates sperm-zona binding and improves efficiency of porcine fertilization in vitro. Biol Reprod 69(3):828–834

McNutt TL, Oldsclarke P, Way AL, Suarez SS, Killian GJ (1994) Effect of follicular or oviductal fluids on movement characteristics of bovine sperm during capacitation in-vitro. J Androl 15(4):328–336

Miller D, Brinkworth M, Iles D (2010) Paternal DNA packaging in spermatozoa: more than the sum of its parts? DNA, histones, protamines and epigenetics. Reproduction 139(2):287–301

Miller D, Ostermeier GC (2006) Spermatozoal RNA: why is it there and what does it do? Gynecol Obstet Fertil 34(9):840–846

Miller D, Ostermeier GC, Krawetz SA (2005) The controversy, potential and roles of spermatozoal RNA. Trends Mol Med 11(4):156–163

Mishima Y, Giraldez AJ, Takeda Y, Fujiwara T, Sakamoto H, Schier AF, Inoue K (2006) Differential regulation of germline mRNAs in soma and germ cells by zebrafish miR-430. Curr Biol 16(21):2135–2142. doi:10.1016/j.cub.2006.08.086

Munson L, Calzada N, Kennedy S, Sorensen TB (1998) Luteinized ovarian cysts in Mediterranean striped dolphins. J Wildl Dis 34(3):656–660

Ostermeier GC, Dix DJ, Miller D, Khatri P, Krawetz SA (2002) Spermatozoal RNA profiles of normal fertile men. Lancet 360(9335):772–777

Ostermeier GC, Miller D, Huntriss JD, Diamond MP, Krawetz SA (2004) Reproductive biology: delivering spermatozoan RNA to the oocyte. Nature 429(6988):154

Overstreet JW, Cooper GW (1975) Reduced sperm motility in the isthmus of the rabbit oviduct. Nature 258:718–719

Pierce GJ, Santos MB, Murphy S, Learmonth JA, Zuur AF, Rogan E, Bustamante P, Caurant F, Lahaye V, Ridoux V, Zegers BN, Mets A, Addink M, Smeenk C, Jauniaux T, Law RJ, Dabin W, Lopez A, Farre JMA, Gonzalez AF, Guerra A, Garcia-Hartmann M, Reid RJ, Moffat CF, Lockyer C, Boon JP (2008) Bioaccumulation of persistent organic pollutants in female common dolphins (*Delphinus delphis*) and harbour porpoises (*Phocoena phocoena*) from western European seas: geographical trends, causal factors and effects on reproduction and mortality. Environ Pollut 153(2):401–415. doi:10.1016/j.envpol.2007.08.019

Rivera RM, Ross JW (2013) Epigenetics in fertilization and preimplantation embryo development. Prog Biophys Mol Biol 113(3):423–432. doi:10.1016/j.pbiomolbio.2013.02.001

Schier AF, Giraldez AJ (2006) MicroRNA function and mechanism: insights from zebra fish. Cold Spring Harb Symp Quant Biol 71:195–203. doi:10.1101/sqb.2006.71.055

Seytanoglu A, Georgiou AS, Sostaric E, Watson PF, Holt WV, Fazeli A (2008) Oviductal cell proteome alterations during the reproductive cycle in pigs. J Proteome Res 7(7):2825–2833. doi:10.1021/pr8000095

Smith TT, Nothnick WB (1997) Role of direct contact between spermatozoa and oviductal epithelial cells in maintaining rabbit sperm viability. Biol Reprod 56(1):83–89

Sonne C (2010) Health effects from long-range transported contaminants in Arctic top predators: an integrated review based on studies of polar bears and relevant model species. Environ Int 36(5):461–491. doi:10.1016/j.envint.2010.03.002

Sonne C, Leifsson PS, Dietz R, Born EW, Letcher RJ, Hyldstrup L, Riget FF, Kirkegaard M, Muir DCG (2006) Xenoendocrine pollutants may reduce size of sexual organs in East Greenland polar bears (*Ursus maritimus*). Environ Sci Technol 40(18):5668–5674. doi:10.1021/es060836n

Sostaric E, Georgiou AS, Wong CH, Moore HD, Watson PF, Holt WV, Fazeli A (2005) Profiling porcine oviductal epithelial cell surface membrane proteins. In: Biology of Reproduction, pp 202–202. 1603 Monroe St, Madison, WI 53711-2021 USA: Soc Study Reproduction

Soubry A (2015) Epigenetic inheritance and evolution: a paternal perspective on dietary influences. Prog Biophys Mol Biol 118(1–2):79–85. doi:10.1016/j.pbiomolbio.2015.02.008

Soubry A, Guo L, Huang Z, Hoyo C, Romanus S, Price T, Murphy SK (2016) Obesity-related DNA methylation at imprinted genes in human sperm: results from the TIEGER study. Clin Epigenetics 8(1):51. doi:10.1186/s13148-016-0217-2

Soubry A, Hoyo C, Jirtle RL, Murphy SK (2014) A paternal environmental legacy: evidence for epigenetic inheritance through the male germ line. BioEssays 36(4):359. doi:10.1002/bies.201300113

Stephane EC, Robert AM (2013) RNA interference in the nucleus: roles for small RNAs in transcription, epigenetics and beyond. Nat Rev Genet 14(2):100. doi:10.1038/nrg3355

Suarez SS, Dai XB, Demott RP, Redfern K, Mirando MA (1992) Movement characteristics of boar sperm obtained from the oviduct or hyperactivated i*n vitro*. J Androl 13:75–80

Suarez SS, Osman RA (1987) Initiation of hyperactivated flagellar bending in mouse sperm within the female reproductive tract. Biol Reprod 36(5):1191–1198

Suarez SS, Pacey AA (2006) Sperm transport in the female reproductive tract. Hum Reprod Update 12(1):23–37. doi:10.1093/humupd/dmi047

Thomas PG, Ignotz GG, Ball BA, Brinsko SP, Currie WB (1995) Effect of coculture with stallion spermatozoa on de novo protein synthesis and secretion by equine oviduct epithelial cells. Am J Vet Res 56(12):1657–1662

Thornburg KL, Shannon J, Thuillier P, Turker MS (2010) In utero life and epigenetic predisposition for disease. Adv Genet 71:57–78. doi:10.1016/B978-0-12-380864-6.00003-1

Vrijens K, Bollati V, Nawrot T (2015) MicroRNAs as potential signatures of environmental exposure or effect: a systematic review. Environ Health Perspect (Online) 123(5):399. doi:10.1289/ehp.1408459

Wells JC, Pomeroy E, Walimbe SR et al (2016) The elevated susceptibility to diabetes in India: an evolutionary perspective. Front Public Health 4:145

Wennemuth G, Carlson AE, Harper AJ, Babcock DF (2003) Bicarbonate actions on flagellar and Ca2+−channel responses: initial events in sperm activation. Development 130(7):1317–1326

Yang X, Zhao Y, Yang X, Kan F (2015) Recombinant hamster oviductin is biologically active and exerts positive effects on sperm functions and sperm-oocyte binding. PLoS One 10(4): e0123003. doi:10.1371/journal.pone.0123003

Yeste M, Holt WV, Bonet S, Rodriguez-Gil JE, Lloyd RE (2014) Viable and morphologically normal boar spermatozoa alter the expression of heat-shock protein genes in oviductal epithelial cells during co-culture in vitro. Mol Reprod Dev 81(9):805–819. doi:10.1002/mrd.22350

Yeste M, Holt WV, Briz M, Bonet S, Lloyd RE (2009a) Boar spermatozoa do not induce changes in heat shock protein gene expression without direct contact with oviductal epithelial cells. Reprod Domest Anim 44:132–132

Yeste M, Lloyd RE, Badia E, Briz M, Bonet S, Holt WV (2009b) Direct contact between boar spermatozoa and porcine oviductal epithelial cell (OEC) cultures is needed for optimal sperm survival in vitro. Anim Reprod Sci 113(1–4):263–278. doi:10.1016/j.anireprosci.2008.08.018

Zhao Y, Srivastava D (2007) A developmental view of microRNA function. Trends Biochem Sci 32(4):189–197. doi:10.1016/j.tibs.2007.02.006

Zhao Y, Yang X, Jia Z, Reid RL, Leclerc P, Kan FWK (2016) Recombinant human oviductin regulates protein tyrosine phosphorylation and acrosome reaction. Reproduction 152(5):561–573. doi:10.1530/REP-16-0177

Epigenetic Influences During the Periconception Period and Assisted Reproduction

Akwasi A. Amoako, Tamer M. Nafee, and Bolarinde Ola

Abstract The periconception period starts 6 months before conception and lasts until the tenth week of gestation. In this chapter, we will focus on epigenetic modifications to DNA and gene expression within this period and during assisted reproduction. There are two critical times during the periconception window when significant epigenetic 'reprogramming' occur: one during gametogenesis and another during the pre-implantation embryonic stage. Furthermore, assisted conception treatments, laboratory protocols and culture media can affect the embryo development and birth weights in laboratory animals. There is, however, an ongoing debate as to whether epigenetic changes in humans, causing embryo maldevelopment, placenta dysfunction and birth defects, result from assisted reproductive technologies or are consequences of pre-existing medical and/or genetic conditions in the parents. The periconception period starts from ovarian folliculogenesis, through resumption of oogenesis, fertilisation, peri-implantation embryo development, embryogenesis until the end of organogenesis. In men, it is the period from spermatogenesis to epididymal sperm storage and fertilisation. Gametes and developing embryos are sensitive to environmental factors during this period, and epigenetic modifications can occur in response to adverse lifestyles and environmental factors. We now know that lifestyle factors such as advanced parentage age, obesity or undernutrition, smoking, excessive alcohol and caffeine intake and recreational drugs used during gamete production and embryogenesis could induce epigenetic alterations, which could impact adversely on pregnancy outcomes and health of the offspring. Furthermore, these can also result in a permanent and irreversible effect in a dose-dependent manner, which can be passed on to the future generations.

A.A. Amoako, PhD, MRCOG
Department of Reproductive Medicine, Central Manchester, University Hospitals NHS Foundation Trust, St Mary's Hospital, Oxford Road, Manchester, UK
e-mail: Akwasi.amoako12@gmail.com

T.M. Nafee, MD, MRCOG • B. Ola, MBBS, FWACS, FRCOG, MD (✉)
Sheffield Teaching Hospitals NHS Foundation Trust, University of Sheffield, Sheffield, UK
e-mail: t.m.nafee@sheffield.ac.uk; Bola.Ola@sth.nhs.uk

A. Fazeli, W.V. Holt (eds.), *Periconception in Physiology and Medicine*,
Advances in Experimental Medicine and Biology 1014,
DOI 10.1007/978-3-319-62414-3_2

Keywords Periconception period • Assisted reproduction technology • Infertility • Epigenetics • Imprinting disorders

1 Introduction

Embryogenesis is a sequence of events that begins with fertilisation to form a single-cell zygote and ends in an organism with more than 200 different cell types. Gene expressions are controlled through two principal mechanisms; the best understood is mediated through the DNA sequence of our genome. The second, not directly related to DNA sequencing, is epigenetic (Nafee et al. 2008). There are at least two critical periods during the periconception window when significant epigenetic 'reprogramming' occurs: one during gametogenesis and another during the pre-implantation embryonic stage. Besides, it is now evident that assisted conception protocols, manipulations and associated culture media can affect the embryo development and birth weights (Dumoulin et al. 2010a; Nelissen et al. 2012; Vergouw et al. 2012). In this chapter, we will focus on how epigenetic modifications to DNA and gene expression within the periconception period and during assisted reproduction affect embryo development. See Fig. 1.

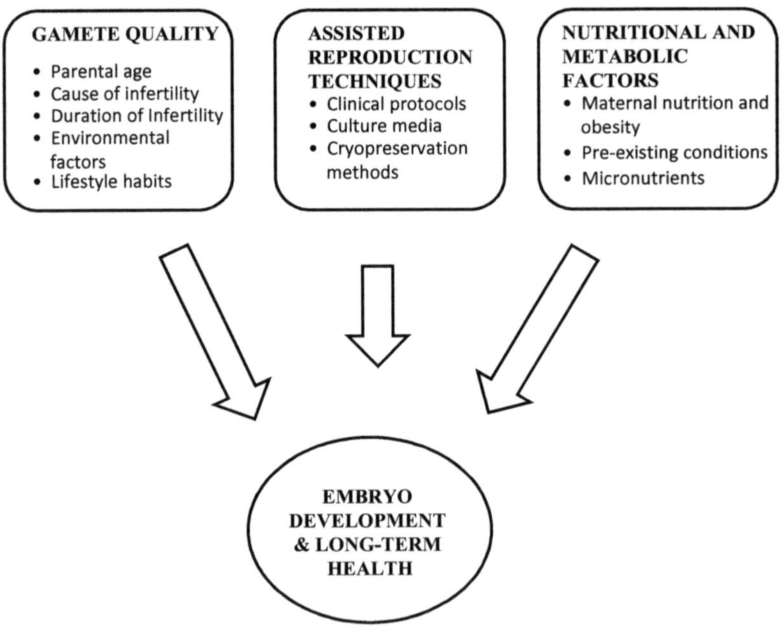

Fig. 1 Periconception epigenetic factors

The periconception period is the 6-month period preceding conception and embryo development to the tenth week of gestation. This period in women is considered to cover ovarian folliculogenesis, resumption of oogenesis, fertilisation, peri-implantation embryo development, embryogenesis and organogenesis. The preconception period in men is the time interval between the start of spermatogenesis to ejaculation and encompasses the processes of spermatogenesis, spermiogenesis, epididymal sperm maturation, epididymal sperm storage and fertilisation. It is estimated that in humans, the onset of spermatogenesis to epididymal sperm maturation typically lasts about 10 weeks (Griffin et al. 2006). Much is known about the adverse impact of periconception lifestyle on gametogenesis, embryo development and reduced reproductive potential (Homan et al. 2007; Hammiche et al. 2011). In women, oogenesis starts during embryonic life and is arrested during first meiotic metaphase until mid-cycle LH surge after puberty. The oocytes are extremely sensitive to environmental factors during this period, and epigenetic modification can occur in response to exposure to adverse lifestyle and environmental factors (Griffin et al. 2006).

There is mounting evidence that periconception lifestyle factors such as advanced parentage age, obesity or undernutrition, smoking, excessive alcohol and caffeine intake and use of recreational drugs during gamete production induce genetic and epigenetic alterations. These changes could impact on pregnancy outcomes and offspring health (Hutt and Albertini 2007; Hutt et al. 2008; Surén et al. 2013). Furthermore, adverse lifestyle factors have been associated with infertility, recurrent implantation failure, miscarriage, intrauterine growth restriction, prematurity and congenital malformations (Steegers-Theunissen et al. 2013). Exposure to adverse lifestyle factors can result in a permanent, irreversible effect in a dose-dependent manner, which can also be passed on to the subsequent generations.

In the last 40 years, enormous advances have been made through assisted reproductive technologies to help infertile couples who would have otherwise remained childless to achieve conception. It is currently estimated that about 4% of all births in the Western world is through in vitro fertilisation (IVF) and intracytoplasmic sperm injection (ICSI) (Sutcliffe and Ludwig 2007; Zegers-Hochschild et al. 2009; De Mouzon et al. 2010).

Assisted reproductive technologies involve in vitro handling and manipulation of gametes and peri-implantation embryos, and concerns about the short-term impact on offspring have been raised since its inception in the 1970s. There is an ongoing debate as to whether techniques and processes involved in assisted reproductive technologies (ART) such as controlled ovarian hyperstimulation, in vitro fertilisation, intracytoplasmic sperm injection, embryo culture, embryo cryopreservation, pre-implantation genetic diagnosis (PGD) and pre-implantation genetic screening (PGS) could alter gamete, embryo and foetal developments.

2 The Impact of Periconception Parental Age on Human Reproduction

2.1 Paternal Age and Fertility

The impact of advanced paternal age on reproductive outcomes remains controversial. Epidemiological studies examining the impact of advanced paternal age on fertilisation, pregnancy rates, miscarriages and live birth have produced conflicting results (Dain et al. 2011). Studies addressing the impact of paternal age on reproductive performance often use semen parameters as the outcome of interest which, in themselves cannot differentiate between fertile and infertile men (Dunson et al. 2004). Although the current quantitative and quality analysis of semen based on current diagnostic techniques may not be different in men of different age groups, advanced paternal age is associated with increased risk of genetic and chromosomal disorders and may account for the poor reproductive function in men of advanced age. Ford et al. (2000) demonstrated that time to conception doubled after the age above 25 years. It is, however, not uncommon for men of advanced age to father offspring and therefore the impact of advanced paternal age on conception is less significant than that of maternal age. Age-related decline in semen volume, sperm motility and morphology has been described. Sperm parameters including semen volume, motility and morphology decrease with age, although a decline in sperm concentration was not demonstrated (Kidd et al. 2001).

Advanced paternal age has been shown to have an adverse impact on conception whether natural (Dunson et al. 2004; Ford et al. 2000; De la Rochebrochard and Thonneau 2003) or through assisted reproductive technologies. A decline in monthly fecundity associated with advanced paternal age has been described, although most of these studies do not account for the impact of maternal factors. Women with male partners aged >45 years have a fivefold increase in the time to achieve natural conception (Hassan and Killick 2003), and this effect was mainly attributed to advanced paternal age after controlling for maternal factors (Hassan and Killick 2003; Stewart and Kim 2001).

Studies exploring the effects of advanced paternal age in the context of assisted reproduction have also produced conflicted results and most often do not account for female factors (Klonoff-Cohen and Natarajan 2004; de La et al. 2006). A significant decrease in fertilisation rate (Luna et al. 2009; Aboulghar et al. 2007), the number of embryo reaching the blastocyst stage and the number of supernumerary embryos available for cryopreservation were reduced in partners of older men undergoing IVF treatment (Luna et al. 2009; Frattarelli et al. 2008). However, another study demonstrated that the live birth rate was not shown to be affected by advanced paternal age (Aboulghar et al. 2007). A retrospective study of 441 couples who underwent 558 oocyte donation cycles did not identify any impact of advanced paternal age on fertilisation and live birth rates (Paulson et al. 2001) suggesting that the effect of paternal age on fertilisation rates and pre-implantation embryo development is small (Dain et al. 2011).

Proponents of an association between advanced paternal age and adverse reproductive outcomes such as infertility and miscarriage have attributed these to the increased risk of epigenetic mutations, DNA mutations and chromosomal aneuploidies. It has been suggested that teratogenic effect of environmental pollutants accumulates as men age and could be linked to increased DNA damage in human spermatozoa (Jafarabadi 2007; Kovac et al. 2013). It is also suggested that DNA damage may result as a consequence of oxidative stress due to increased free oxygen radicals in the semen of older men (Cocuzza et al. 2008). Paternal age was shown in one study to be associated with increased risks of spontaneous abortions and Down's syndrome. Other studies have however found little or no association between advanced paternal age and risk of spontaneous abortion after controlling for maternal age (Kazaura and Lie (2002); Kleinhaus et al. 2006; Maconochie et al. 2007).

2.2 *Maternal Age*

Maternal age remains the most significant factor influencing conception and offspring health in human reproduction. The natural age-related decline in female fertility is a well-established association with natural conceptions and assisted reproductive technologies. After the age of 35 years, female fertility declines, and time to conception increases. The decline in female fertility arises as a result of ovarian ageing associated with increased maternal age (Broekmans et al. 2004). A decline in ovarian function occurs towards the end of women's reproductive life due to exhaustion of primordial follicles (Hansen et al. 2008; Broekmans et al. 2009; Johnson et al. 2006).

The decline in ovarian reserve means that as women age, fewer follicles are available for selection and follicles of poor quality which would otherwise undergo atresia will be selected. Oocyte quality also decreases with age as a result of cumulative damage to the oocyte, spindle fibres and declining quality of its surrounding granulosa cells (Warburton 2005).

Women of advanced age with fertility problems resort to assisted reproductive technologies for help. Although assisted reproductive technologies offer hope to these women by decreasing time to conception, live birth rate decreases with all forms of assisted reproduction as maternal age increases (Dovey et al. 2008; Harris et al. 2010; Gunby et al. 2010; Leridon 2004).

Reports from various studies and national databases show that the current live birth rate for women aged >42 years old undergoing IVF treatment is in the region of 5% (Gunby et al. 2011; Hourvitz et al. 2009).

Advanced maternal age is associated with an increased risk of pregnancy loss in addition to infertility. Effects of age are largely attributed to the increased risk of aneuploidy as a result of ovarian ageing which results in an increased rate of chromosomally abnormal conceptions. In a study of oocytes that failed to fertilise or cleave, chromosomal aneuploidy resulting from meiotic errors increased with age, although the study suffered from a lack of control (Pellestor et al. 2003). It would have been useful if there was a control group of fertilised oocytes from women undergoing PGD; in order to investigate whether aneuploidies also increased more or less with maternal age in this group. For example, in another study, aneuploidies were significantly more common chromosomal abnormalities found in recurrent miscarriages compared with control group of couples undergoing PGD for sex-linked diseases (Rubio et al. 2003).

3 The Impact of Periconception Parental Obesity, Nutrition and Smoking Status on Conception

3.1 Paternal Obesity and Nutrition

The current global obesity epidemic has had some impact on semen quality and male reproductive potential. There is no agreement on the link between body mass index (BMI) and sperm concentration or total sperm count (Paasch et al. 2010; Aggerholm et al. 2008; Duits et al. 2010; MacDonald et al. 2010). Nevertheless, a trend towards paternal obesity and decreased sperm concentration resulting in infertility and reproductive failure has been observed. An updated systematic review by Sermondade et al. (2012) showed that obese men are more likely to present with oligozoospermia or azoospermia, compared with normal-weight men. The decline in sperm quality in obese men may be attributable to alterations in reproductive hormones such as oestrogen, serum testosterone and sex hormone binding globulin levels (Pasquali 2006; Qin et al. 2007; Pasquali et al. 2007; Hammoud et al. 2008a).

A recent systematic review and meta-analysis exploring the impact of BMI on sperm count found an association between obesity and abnormal semen parameters. Sermondade et al. (2013) compared sperm parameters of normal men with underweight, overweight, obese and morbidly obese men. The odds ratios (95% CI) for oligozoospermia or azoospermia were 1.15 (0.93–1.43), 1.11 (1.01–1.21), 1.28 (1.06–1.55) and 2.04 (1.59–2.62), respectively. The conclusion was that overweight and obesity were associated with an increased prevalence of azoospermia or oligozoospermia (Sermondade et al. 2012). Epidemiological studies have also revealed that women who have overweight, obese male partners or heavy cigarette smokers have elevated risk of infertility (Sallmén et al. 2006; Nguyen et al. 2007; Ramlau-Hansen et al. 2007a, 2007b).

A systematic review and meta-analysis on the effect of obesity on reproductive potential concluded that obese men were more prone to infertility (OR = 1.66, 95% CI 1.53–1.79), reduced live birth rate after ART (OR = 0.65, 95% CI 0.44–0.97) and a 10% increase in miscarriages (Campbell et al. 2015).

Possible explanations for the poor performance of spermatozoa from obese men have included low mitochondrial energy potential, increased DNA fragmentation, abnormal morphology and altered DNA methylation, which reduces their fertility potential (Soubry et al. 2013).

3.2 *Maternal Obesity and Nutrition*

Obesity in women of reproductive age has become prevalent worldwide, and the negative consequence on reduced fertility is well recognised. Obesity is associated with ovulatory disorders, menstrual disorders, increased time to conception, infertility, miscarriage and congenital abnormalities (Pasquali et al. 2007; Practice Committee of American Society for Reproductive 2008; Brewer and Balen 2010; Grindler and Moley 2013). Oocytes depend on their mitochondria for energy and cellular homoeostasis. Environmental and nutritional factors can adversely affect mitochondrial DNA copy number and function. These explain why obesity associated reproductive disorders can result from dysfunction of oocyte mitochondria leading to poor oocyte quality (Grindler and Moley 2013). Furthermore, obese women undergoing fertility treatment have poorer outcomes, have lower pregnancy rates and more likely to miscarry (Maheshwari et al. 2007). In IVF cycles, obese women require much higher doses of gonadotrophins, produce poor-quality oocytes and have reduced success rates compared to normal-weight controls (Maheshwari et al. 2007; Dokras et al. 2006; Erel and Senturk 2009; Depalo et al. 2011; Luke et al. 2011). Furthermore, gestational diabetes and pre-eclampsia are more common in obese women compared with normal-weight controls (Maheshwari et al. 2007).

3.3 *Maternal Micronutrients: Folate and Vitamin D*

The role of preconception folic acid supplementation in the prevention of neural tube defects has been established since 1991, through a double-blind, randomised, controlled trial, performed across seven countries (MRC Vitamin Study Research Group – 1991). Since then, many other studies in human populations demonstrated a beneficial effect of periconceptional folic acid in the prevention of preterm labour (Li et al. 2014); pregnancy-induced hypertension (Yang et al. 2016), small for gestational age babies (Hodgetts et al. 2015) and the growth velocity of the human embryonic cerebellum (Koning and Steegers-Theunissen 2015). In a large study of the association between maternal plasma folate during pregnancy and epigenome-wide DNA methylation in 1988 newborns from two European cohorts, 48 genes were significantly altered in babies with maternal serum folate levels (Joubert et al. 2016), implying an epigenetic mechanism for the action of folate. There is increasing attention on non-skeletal benefits mediated by periconceptional Vitamin D supplements. The mode of action is not fully understood, but the evidence is mounting that these effects are mediated via epigenetic controls (Hossein-nezhad and Holick 2012).

3.4 The Impact of Lifestyles on Fertility

3.4.1 Effects of Maternal Periconception Use of Caffeine, Alcohol and Smoking on Fertility

Evidence from some studies has shown that maternal caffeine use in the periconception period negatively impacts on fertility (Klonoff-Cohen et al. 2002; Homan et al. 2007; Anderson et al. 2010a, b; Wesselink et al. 2016) in a dose-dependent manner. The current recommendation is that women attempting to conceive should limit caffeine intake to 1–2 cups of coffee per day (Homan et al. 2007). The exact mechanism by which caffeine impacts on fertility is unclear, but it is purported to affect ovulation and corpus luteum function and to alter female reproductive hormone levels. Studies have suggested a relationship between caffeine intake and increased time to conception (Anderson et al. 2010a, b). Increased caffeine consumption in the periconception period decreases implantation and live birth rates and increases miscarriages in women undergoing assisted reproduction (Klonoff-Cohen et al. 2002).

Maternal alcohol use also has adverse consequences on fertility, and the current evidence recommends that women planning to conceive should stop drinking alcohol (Anderson et al. 2010a, b). There are no dose-effect studies, and there is no recommended safe limit of alcohol intake in the periconception period. The mechanism by which alcohol impairs fertility is still unclear. High maternal alcohol consumption reduces chances of live birth and increases the risk of miscarriage in both spontaneous and assisted conceptions (Klonoff-Cohen et al. 2002).

There is substantial research evidence that maternal smoking adversely affects female fertility. Female smokers take a longer time to conceive (Hull et al. 2000) and are more likely to be infertile compared to non-smokers (Homan et al. 2007). The impact of smoking on female fertility is dose dependent, and heavy smokers are more likely to be affected although low levels of smoking can have a significant adverse effect on female fertility. Female smoking accelerates loss of ovarian function (ESHRE Task Force on Ethics and Law 2010) and decreases endometrial receptivity (Soares et al. 2007). It also increases spontaneous miscarriage in both natural and assisted conception cycles (Winter et al. 2002) and increases the risk of ectopic pregnancies (Anderson et al. 2010a, b). In assisted reproduction treatment, maternal smoking has been shown to affect success rates significantly by causing a 50% reduction in implantation rate (Cooper and Moley 2008) compared to non-smokers (Waylen et al. 2009; Meeker and Benedict 2013). Maternal smoking is also associated with increased risk of intrauterine growth restriction, low birth weight, preterm birth (McCowan and Horgan 2009; Juárez and Merlo 2013) and birth defects (Hackshaw et al. 2011).

3.4.2 Effects of Paternal Periconception Use of Caffeine, Alcohol and Smoking on Fertility

There is good evidence that paternal preconception smoking adversely affects male fertility (Homan et al. 2007). The impact of smoking on male fertility is dose dependent, although light smoking is still associated with impaired male fertility (Ramlau-Hansen et al. 2007a, b). Men who smoke have lower sperm concentration (Aryanpur et al. 2011), have reduced sperm motility and viability (Gaur et al. 2010) and have a higher proportion of abnormal sperm (Gaur et al. 2010; Ramlau-Hansen et al. 2007a, b). Besides, smoking increased sperm aneuploidy and, overall, reduced fertilising capacity (Said et al. 2005; Sofikitis et al. 2000). Smoking increases oxidative damage to sperm DNA by decreasing seminal fluid antioxidant activity (Pasqualotto et al. 2008). Exposure to the toxicants in cigarette smoke has been shown to increase free oxygen radicals in semen, causing oxidative damage to sperm plasma membranes and sperm DNA integrity and inducing genetic and epigenetic abnormalities. The integrity of the paternal genome plays a crucial role in successful fertilisation and normal embryo development. Damage to the paternal DNA by tobacco smoke results in the transmission of abnormal genetic information into the oocyte during fertilisation, which results in abnormal embryo development. Consequently, this may result in abnormal pregnancy outcomes such as recurrent implantation failure, miscarriage and congenital malformations. Emerging evidence suggests that the negative impact of environmental toxicants induces permanent heritable genetic and epigenetic alterations in spermatozoa which can be passed on to subsequent generations. Individual susceptibility to the negative effect of toxicants in cigarette smoke depends on other lifestyle factors, genetics and the duration and dose of exposure (Marczylo et al. 2012).

Paternal preconception alcohol consumption is associated with deterioration of semen parameters in a dose-dependent manner (Gaur et al. 2010; Muthusami and Chinnaswamy 2005; Jensen et al. 2014) and reduces live birth rates as well as increasing miscarriage rate in assisted reproduction treatments (Klonoff-Cohen et al. 2003). Moderate alcohol consumption does not appear to impair semen parameters significantly (Jensen et al. 2014). However, heavy and chronic alcohol intoxication induces oligo-, astheno- and teratozoospermia and eventually azoospermia (Sermondade et al. 2010).

Paternal preconception caffeine intake seems to confer some protection to sperm by increasing sperm motility both as dietary and in vitro supplement (Sobreiro et al. 2005), although other studies have not demonstrated any beneficial effect (Jensen et al. 2010; Klonoff-Cohen et al. 2002). Emerging evidence suggests that high caffeine consumption may even have harmful effects on sperm and Sertoli cells by decreasing the antioxidant capacity and increasing oxidative damage (Dias et al. 2015).

4 Risk of Assisted Reproductive Technologies

The short-term impact of assisted reproductive technologies has been a subject of intense debate for the last three decades. So far no substantial evidence exists to suggest that the use of assisted reproduction technologies is associated with short-term adverse consequences such as low implantation rate, decreased conception rates or increased risk of miscarriages. There are, however, reports of increased incidence of birth defects, chromosomal abnormalities (Bonduelle et al. 2002), prematurity, intrauterine growth retardation (Schieve et al. 2004; Schieve et al. 2007; Romundstad et al. 2008), epigenetic changes and imprinting disorders (Khosla et al. 2001a, b). However, the unanswered question is whether these are due to the underlying cause of the couple's infertility or due to assisted reproductive technologies. There is an argument that parental background during the periconception period, as well as the underline causative factors in infertility, may be responsible for these changes and not the techniques used (Myklestad et al. 2012; Belva et al. 2012).

Controlled ovarian hyperstimulation plays a major role in ART, but concerns have been raised about its detrimental effects on oocyte maturation, embryo quality, endometrial receptivity, post-implantation embryo loss and perinatal outcomes. Ovulation induction using supraphysiological doses of gonadotrophins creates an artificial oviductal and endometrial environment as well as alteration of oocyte maturation (Santos et al. 2010; Ozturk et al. 2016). It has been proposed that controlled ovarian hyperstimulation may alter gene expression related to oocyte and early embryo development (Ozturk et al. 2016).

In IVF/ICSI cycles, gametes and embryos are subjected to artificial culture media which alters the hormonal and chemical environment with potential consequences such as alterations in DNA methylation, mRNA-mediated abnormal expression of genes and epigenetic modifications. These were some of the reasons used to explain findings in studies showing that different culture media caused significant changes in mean singleton birth weights (Dumoulin et al. 2010b) and placental weight (Eskild et al. 2013). Furthermore, extended embryo culture to the blastocyst stage was shown to increase the incidence of large for gestational age babies (Mäkinen et al. 2013).

In mammals, in vitro embryo culture was associated with aberrant genomic imprinting, altered intrauterine growth pattern and defective placentation (Khosla et al. 2001a, b). Epigenetic alterations of some imprinted genes were shown to vary with the method of fertilisation and the type and constituents of the culture medium used and to be tissue specific in bovine embryo. In vitro culture of mouse embryos up to the blastocyst stage, using some commercially available sequential media, resulted in a pronounced permanent shift in the expression of some non-imprinted genes, despite an apparently normal growth, development and morphometric measurements of the animals born (Morgan et al. 2008). The reports of susceptibility of such mice to hypertension despite their normal post-natal growth raises concerns (Watkins et al. 2007, 2008). So also, the potential trans-generational persistence of some of these adverse effects into the progeny of the next generation after normal mating is possible (Mahsoudi et al. 2007). Although these observations were in animal studies, they raise an alarm about the possible end-points for babies born to current assisted conception techniques.

Intracytoplasmic sperm injection (ICSI) also raises some issues of concern. Genetic and chromosomal abnormalities are commonly found in spermatozoa from men with severe male factor infertility. As ICSI bypasses the natural selection of sperm in spontaneous conceptions and conventional IVF, there is a risk of injecting sperm with genetic or chromosomal disorders which may impact on embryo development or cause transmission of chromosomal abnormality in the case of sperm aneuploidy (Bonduelle et al. 2002).

5 Consequences of ART

5.1 Perinatal Outcome

Several studies and data from national and international birth registries have shown an increased risk of adverse perinatal outcome in singleton infants conceived by IVF and ICSI (Schieve et al. 2004, 2007). The offspring are more prone to intrauterine growth restriction, low birth weight, preterm delivery and perinatal deaths compared with infants born to fertile couples (Helmerhorst et al. 2004; Jackson et al. 2004; Sutcliffe and Ludwig 2007). Iwahata and colleagues (2015) compared the closed and open vitrification systems in a retrospective study and showed no significant differences in post-implantation embryo development, gestational age, birth weight, sex ratio, Apgar scores or neonatal congenital anomalies.

Other researchers have found an increase in the rate of adverse outcomes in IVF singleton pregnancies. These include an increase in the incidence of low birth weight (LBW) (Schieve et al. 2002), preterm birth (PTB), hospitalisation during pregnancy and perinatal mortality (Gissler et al. 1995). Others include pregnancy-induced hypertension, placenta praevia (Tanbo et al. 1995; Romundstad et al. 2006) and, more recently, stillbirth (Wisborg et al. 2010; Zhu et al. 2006). A recent and comprehensive meta-analysis used pooled data from 17 studies to determine the rate of obstetric complications in IVF singletons and found an association with adverse perinatal outcome, with an increased risk of PTB (RR 1.84, 95% CI 1.54, 2.21) and LBW (RR 1.60, 95% CI 1.29, 1.98) (data from McDonald et al. 2009). Therefore, it appears that the underlying causes of infertility, ovarian stimulation or the IVF method are attributable to adverse outcomes in various populations.

Other studies have, however, refuted these claims (Belva et al. 2012) and associated the increase in adverse perinatal outcomes to maternal and other factors rather than the assisted reproductive technologies (Myklestad et al. 2012). A recent systematic review and meta-analysis found an association between length of time to conception and increased risk of preterm labour in women who conceived spontaneously: pooled crude odds ratio (OR): 1.38 (95% CI: 1.25–1.54) (Messerlian et al. 2012).

5.2 Placental Dysfunction

The increased risk of adverse perinatal outcomes in ART-conceived pregnancies is largely attributable to placental dysfunction resulting from epigenetic alterations due to in vitro gamete and embryo handling. Although the epigenetic marks of the developing trophoblast are very distinct from those of the inner cell mass and the developing mammalian embryo (Santos and Dean 2004; Santos and Dean 2006), it may be epigenetically as vulnerable. Human trophoblastic cell migration and invasion are epigenetically regulated processes (Rahnama et al. 2006). Loss of imprinting, altered epigenetic marks and gene expression patterns were observed in small for gestational age (SGA) babies (Guo et al. 2008). Placentae of babies born to in vitro conception demonstrated altered DNA methylation patterns and gene expression profiles when compared to those resulting from spontaneous conception (Katari et al. 2009) in humans.

5.3 Birth Defects

Children from pregnancies occurring, either spontaneously or following induced folliculogenesis, after a period of subfertility, are also at higher risk of congenital malformations (Sutcliffe and Ludwig 2007). Chung et al. (2006) found no effect of aetiology of subfertility, the type or dose of stimulation drugs, use of embryo manipulation techniques or embryo quality on perinatal outcomes. The evidence of an increased risk of congenital anomalies and assisted conception has however always been contradictory (Hart and Norman 2013a, b). Anthony et al. (2002) found no significant associations between IVF and an increased risk of overall congenital malformation of different types and severity. However, an increased risk of congenital abnormalities in singleton IVF boys and congenital heart abnormalities in girls conceived via alternative ART procedures was demonstrated (Klemetti et al. 2005), in addition to an association between musculoskeletal defects and intrauterine insemination (IUI) (Olson et al. 2005). Pooled data from some studies has also found a significant increase in 45 cohort studies comprising 92,671 ART babies compared with 3,870,760 naturally conceived infants R.R. 1.32 (95% CI 1.24–1.42) (Hansen et al. 2013). The overall risk of anomalies could have arisen from either IVF procedure or the super-ovulation medications or intrinsic factors associated with infertility (Ludwig 2009; Hammoud et al. 2008a).

More recently, Kissin et al. (2015) showed that approximately 0.8% of ART children (P for trend 0.19) had autism; which was lower following ART in parents diagnosed with unexplained subfertility, compared with ART for other causes of infertility (adjusted hazard risk ratio [aHRR]; 95% CI: 0.38; 0.15–0.94). In the same study, the incidence of autism was higher when ICSI, rather than IVF, was used (aHRR 1.65; 1.08–2.52). Heisey et al. (2015), in a large retrospective cohort study, found that relative risk of congenital malformations was 1.43 (95% CI 1.19–1.72) for singleton infants conceived through ART compared with natural conceptions. The particular ART-associated defects were hypospadias, patent ductus arteriosus and obstructive uropathies.

In a case-control study involving 380 cases comparing foetal anorectal malformations in IVF/ICSI and fertile patients, Wijers et al. (2015) found increased risks of anorectal malformations in singleton ART pregnancies following IVF (OR = 2.4; 95% CI = 1.0–5.90 and ICSI (OR = 4.2; CI = 1.9–8.9), respectively. Anorectal malformations were also more common in offspring of subfertile parents conceived by IVF treatment, compared with those of subfertile parents who eventually conceived naturally (OR 3.2; CI = 1.4–7.2).

Some studies associate congenital anomalies with assisted conception treatments; others do not. Seggers et al. (2015) studied offspring with congenital anomalies from a registry-based study in Northern Netherland. A total of 4185 malformed cases recorded in fertile couples and 340 in subfertile couples. Of these, 139 had conceived after IVF/ICSI and 201 naturally after >12 months. Seggers et al. (2015) linked subfertility to an increased risk of ventral wall defects (adjusted OR [aOR] 2.43, 95% CI 1.05–5.62) and penoscrotal hypospadias (aOR 9.83, 95% CI 3.58–27.04). Other associations were methylation defects causing imprinting disorders (aOR 13.49, 95% CI 2.93–62.06) and right outflow tract obstruction (aOR 1.77, 95% CI 1.06–2.97). In the same study (Seggers et al. 2015), in vitro fertilisation/ICSI was associated with an increased risk of polydactyly (OR 4.83, 95% CI 1.39–16.77) and more specifically polydactyly of the hands (OR 5.02, 95% CI 1.43–17.65).

A contrary view is supported by Chen et al. (2015) who conducted a large retrospective study, including 2060 live-born infants, corresponding to 1622 frozen-thawed (FET) cycles. There were 587 live-born infants from long protocol stimulation (LPS) (including 458 FET cycles) and 1257 live-born infants from the short protocol (SPS) (including 984 FET cycles) and 216 live-born infants from mild ovarian stimulation (including 180 FET cycles). Chen et al. (2015) found that multiple pregnancies, gestational age, baby weight and length and early neonatal deaths were comparable in all groups. Live-birth defects in the LPS group (1.02%) and SPS (0.64%) were not significantly different compared to the mild ovarian stimulation group (0.46%). Congenital malformations were, however, significantly increased for the infertility-duration and multiple pregnancies (the adjusted OR were 1.16 (95% CI =1.1–1.335 and OR 3.89 (95% CI = 1.18–12.89), respectively. No links were found between congenital birth defects and types of ovarian stimulation regimens, BMI, maternal age, parity, insemination method or infant gender.

5.3.1 Epigenetic Impact of Assisted Reproductive Technologies

Major epigenetic reprogramming in human reproduction occurs during gametogenesis and peri-implantation embryo development. The periconception period is, therefore, a critical stage for epigenetic events, and epigenetic disruptions may occur during gametogenesis, fertilisation and early embryo development (Reik and

Walter 2001; Strachan and Read 2003). Biological and lifestyle factors can, therefore, influence epigenetic reprogramming of offspring and induce permanent heritable modifications in gene expression without altering DNA sequence (Goldberg et al. 2007; Marczylo et al. 2012). On fertilisation, this extensive reprogramming effectively combines the two highly differentiated, transcriptionally silent gametes into a totipotent embryo and then extends further beyond embryonic genome activation to lay the cell lineage-specific imprints of the differentiating blastocyst (Fulka et al. 2004; Fulka et al. 2008). The process involves extensive modifications involving several inter-dependent layers of epigenetic controls, like covalent histone modifications, chromatin remodelling, DNA methylation and post-translational inhibition through microRNAs (Tang et al. 2007; Yang et al. 2008). The integrity of these processes is shown, in mammalian models, to be crucial for a smooth maternal-to-zygotic transition, through a timely embryonic genome activation, appropriate gene expression and, hence, proper embryo differentiation and development (Bell et al. 2008; Santos et al. 2010; Schier 2007; Shi and Wu 2009; Tang et al. 2007). It is during this extremely labile phase that the gametes and embryos are subjected to the unnatural stresses of ovarian hyperstimulation, in vitro culture and maturation and micromanipulation.

Exposure of gametes and embryos in vitro to artificial conditions and non-physiological media may lead to epigenetic mutations. Controlled ovarian hyperstimulation has been associated with increased risk of epimutations. Oocytes retrieved from mice and humans following controlled ovarian hyperstimulation have altered methylation of imprinting genes (Sato et al. 2007; Fauque et al. 2007). IVF culture media can also modulate DNA methylation and epigenetic mutation. Fauque et al. (2007) showed that culture media can influence DNA methylation and kinetics of cleavage stage embryos. Animal studies demonstrate an impact of assisted reproductive techniques on various aspects of embryonic DNA methylation, histone modifications, miRNA populations and gene expression studies (Fauque et al. 2007; Mtango et al. 2009; Qiao et al. 2010). Fewer than 50% of IVF embryos reach the blastocyst stage of development, with many of these unable to sustain development following embryo replacement (Betts and Madan 2008). It has been clearly documented that aberrant methylation and histone deacetylation during meiosis (Akiyama et al. 2006) result in embryonic abnormalities, aneuploidies and developmental arrest. The possibility that in vitro conception may stretch the epigenetic plasticity of some human embryos beyond their adaptive capacity cannot be ignored.

5.3.2 ART and Imprinting Disorders

Imprinting disorders like Beckwith-Wiedemann and Angelman syndromes appear to be increased in offspring from assisted reproductive technologies (Maher 2005; Sutcliffe et al. 2006; Amor and Halliday 2008). Most studies reporting an association between ART and imprinting disorders do not, however, control for parental genetic and epigenetic factors like paternal origins of chromosome 15 deletions or maternal disomy 15, infertility, obesity and high family income (Chang et al. 2005). Other studies have not found any associations between the incidence of either the Angelman or Prader-Willi syndrome and IVF or ICSI treatments when corrected for parental subfertility (Vermeiden and Bernardus 2013; Bowdin et al. 2007; Sutcliffe et al. 2006). This raises the possibility that the association is more likely to be related to parental subfertility rather than ART (Ludwig et al. 2005). Altered DNA methylation in sperm from men with oligozoospermia, teratozoospermia or oligoasthenoteratozoospermia has been found, (Boissonnas et al. 2010; Marques et al. 2004) and preliminary data suggest that oligospermic patients may carry a risk of transmitting primary incorrect imprints to the offspring (Kobayashi et al. 2009).

5.4 Mitochondrial Replacement Therapy

There are many uncertainties about the genetic and epigenetic risks to babies born after mitochondrial replacement therapy. There are ethical and medical concerns relating to the offspring conceived after mitochondrial DNA replacement therapy (MRT); but many of these have not yet been studied. Human mitochondria contains only about 37 genes, and these contribute less than 0.1% of genomic DNA, yet thousands of women all over the world are at risk of transmitting mitochondrial diseases to their children, with most babies dying shortly after birth. For example, it was estimated that 1 in 4000 children were born in the USA with this condition (Schaefer et al. 2004). Recently, mitochondrial DNA replacement therapy (MRT), commonly referred to as 3-parent IVF, has become accepted in the UK and declared permissible ethically by the FDA.

Although abnormal mutations have been identified in 13 mtDNA genes, which are linked to mitochondrial diseases, it is however less clear how specific genetic defects are exactly linked to disorders of tissues, organs or systems (Koopman et al. 2012). Furthermore, we do not yet know of possible negative impacts of MRT on progressive enhancements of mitochondrial-nuclear allelic interactions, which are evolutionary responses to natural selection (Reinhardt et al. 2013). It is also suspected that mitochondrial replacement may have unknown effects on subsequent epigenetic programming during embryo and foetal development (Ishii 2014). Admittedly, however, it may take a long time before the risk-harm balances of these concerns are resolved by scientific studies.

5.5 Recreational Drugs (Licit and Illicit Drugs of Abuse)

Recreational drugs form a large spectrum, including alcohol and cigarette smoking which have already been covered. Others include amphetamine-group stimulants (e.g. amphetamine and methamphetamine, fenethylline, methcathinone, methylphenidate, ephedrine, pseudoephedrine, 'ecstasy' and other hallucinogens), tranquilisers, opioids (heroin, morphine, codeine, oxycodone, hydrocodone and methadone), cocaine and cannabis. The use of recreational drugs is a worldwide problem. This is illustrated by the report of the US Department of Health and Human Services agency, the Substance Abuse and Mental Health Services Administration (SAMHSA 2015), which reported that 10.2% of Americans aged over 12 years had used illicit drug in 2014. This figure surpassed previous years since 2002 and was mainly accounted for by cannabis and non-medical pain relievers. Recreational drugs may lead to maternal, foetal and neonatal complications, depending on timing, dosing and duration of use. Teratogenic effects may occur if the foetus is exposed in the periconception period. However, long-term disorders of foetal growth and neuro-developmental impairment can also occur, as well as neonatal sequelae like congenital anomalies, sudden infant death syndrome, neonatal abstinence syndrome, respiratory distress syndrome and neurobehavioral changes (Miller and Hyatt 1992; Huestis and Choo 2002). Oxidative stress and genetic and epigenetic mechanisms have recently been implicated in affected offspring of addicted mothers (Kovatsi et al. 2011; Neri et al. 2015). 'Transgenerational epigenetics' have also been proposed as the cause of developmental abnormalities, impairment in learning and memory and attention deficit associated with periconceptional use of recreational drugs (Neri et al. 2015).

Huestis and Choo (2002) showed that over 75% of exposed infants developed major medical problems compared with 27% of unexposed. They also showed that 20% of exposed babies were born prematurely, compared with 6% of unexposed neonates prematurely. Antenatal use of amphetamine may cause foetal demise, prematurity, low birth weight, growth retardation, birth defects and development disorders. Neonates may also be jittery and hypersensitive to touch. Otero et al. (2004) found poor social adjustment, cognitive deficits and learning disabilities in older children manifesting long-term sequelae of antenatal methamphetamine use. Periconceptional and prenatal abuse of cocaine or marijuana has been linked with miscarriage, premature birth, low birth weight and decreased head circumference (Phibbs et al. 1991 and Zuckerman et al. 1989). In a study of 125 children followed up from 3–6 months until their fourth year, it was demonstrated that long-term sequelae of prenatal cocaine abuse include irritability, attention deficit and behavioural problems (Vogel 1997).

6 Conclusion

In conclusion, the period from 24 weeks preceding a pregnancy to 10 weeks after conception has significant impacts on human reproduction. This periconception period includes gametogenesis, gamete maturation, fertilisation, peri-implantation embryo development, embryogenesis and organogenesis. Advanced parentage age, lifestyle habits such as obesity or under nutrition, smoking, excessive alcohol and caffeine intake and use of recreational drugs can induce genetic and epigenetic alterations, which could cause infertility, implantation failure, recurrent miscarriages and adverse pregnancy and perinatal outcomes. The short-term impact of assisted reproductive technologies has been a subject of intense debate for the last three decades. So far, no substantial evidence exists to suggest that the use of assisted reproduction technologies is associated with short-term adverse consequences such as low implantation rate, decreased conception rates or increased risk of miscarriages. Although there are reports of increased incidence of birth defects, chromosomal abnormalities, prematurity, intrauterine growth retardation, epigenetic changes and imprinting disorders, it is unclear whether these are due to the underlying cause of the couple's infertility or side-effects of assisted reproductive technologies.

References

Aboulghar M, Mansour R, Al-Inany H, Abou-Setta AM, Aboulghar M, Mourad L, Serour G (2007) Paternal age and outcome of intracytoplasmic sperm injection. Reprod BioMed Online 14(5):588–592

Aggerholm AS, Thulstrup AM, Toft G, Ramlau-Hansen CH, Bonde JP (2008) Is overweight a risk factor for reduced semen quality and altered serum sex hormone profile? Fertil Steril 90(3):619–626

Akiyama T, Nagata M, Aoki F (2006) Inadequate histone deacetylation during oocyte meiosis causes aneuploidy and embryo death in mice. Proc Natl Acad Sci 103(19):7339–7344

Amor DJ, Halliday J (2008) A review of known imprinting syndromes and their association with assisted reproduction technologies. Hum Reprod 23(12):2826–2834

Anderson K, Nisenblat V, Norman R (2010a) Lifestyle factors in people seeking infertility treatment–a review. Aust N Z J Obstet Gynaecol 50(1):8–20

Anderson K, Norman RJ, Middleton P (2010b) Preconception lifestyle advice for people with subfertility. Cochrane Database Syst Rev 14(4)

Anthony S, Buitendijk SE, Dorrepaal CA, Lindner K, Braat DDM, Den Ouden AL (2002) Congenital malformations in 4224 children conceived after IVF. Hum Reprod 17(8):2089–2095

Aryanpur M, Tarahomi M, Sharifi H, Heydari G, Hessami Z, Akhoundi M, Masjedi MR (2011) Comparison of spermatozoa quality in male smokers and nonsmokers of Iranian infertile couples. Int J Fertil Steril 5(3):152–157

Bell CE, Calder MD, Watson AJ (2008) Genomic RNA profiling and the programme controlling preimplantation mammalian development. Mol Hum Reprod 14(12):691–701

Belva F, Roelants M, De Schepper J, Roseboom TJ, Bonduelle M, Devroey P, Painter RC (2012) Blood pressure in ICSI-conceived adolescents. Hum Reprod 27(10):3100–3108

Betts DH, Madan P (2008) Permanent embryo arrest: molecular and cellular concepts. MolHumReprod 14(8):445–453

Boissonnas CC, El Abdalaoui H, Haelewyn V, Fauque P, Dupont JM, Gut I et al (2010) Specific epigenetic alterations of IGF2-H19 locus in spermatozoa from infertile men. Eur J Hum Genet 18(1):73–80

Bonduelle M, Van Assche E, Joris H, Keymolen K, Devroey P, Van Steirteghem A, Liebaers I (2002) Prenatal testing in ICSI pregnancies: incidence of chromosomal anomalies in 1586 karyotypes and relation to sperm parameters. Hum Reprod 17(10):2600–2614

Bowdin S, Allen C, Kirby G, Brueton L, Afnan M, Barratt C, …, Reardon W (2007) A survey of assisted reproductive technology births and imprinting disorders. Hum Reprod 22(12): 3237–3240

Brewer CJ, Balen AH (2010) The adverse effects of obesity on conception and implantation. Reproduction 140(3):347–364

Broekmans FJ, Faddy MJ, Scheffer G, te Velde ER (2004) Antral follicle counts are related to age at natural fertility loss and age at menopause. Menopause 11(6, Part 1 of 2):607–614

Broekmans FJ, Soules MR, Fauser BC (2009) Ovarian aging: mechanisms and clinical consequences. Endocr Rev 30(5):465–493

Campbell JM, Lane M, Owens JA, Bakos HW (2015) Paternal obesity negatively affects male fertility and assisted reproduction outcomes: a systematic review and meta-analysis. Reprod BioMed Online 31(5):593–604

Chang AS, Moley KH, Wangler M, Feinberg AP, DeBaun MR (2005) Association between Beckwith-Wiedemann syndrome and assisted reproductive technology: a case series of 19 patients. Fertil Steril 83(2):349–354

Chen H, Wang Y, Lyu Q, Ai A, Fu Y, Tian H, …, Kuang Y (2015) Comparison of live-birth defects after luteal-phase ovarian stimulation vs. conventional ovarian stimulation for in vitro fertilization and vitrified embryo transfer cycles. Fertil Steril 103(5):1194–1201

Chung K, Coutifaris C, Chalian R, Lin K, Ratcliffe SJ, Castelbaum AJ, Freedman MF, Barnhart KT (2006) Factors influencing adverse perinatal outcomes in pregnancies achieved through use of in vitro fertilization. Fertil Steril 86(6):1634–1641

Cocuzza M, Athayde KS, Agarwal A, Sharma R, Pagani R, Lucon AM, …, Hallak J (2008) Age-related increase of reactive oxygen species in neat semen in healthy fertile men. Urology 71(3):490–494

Cooper AR, Moley KH (2008, March) Maternal tobacco use and its preimplantation effects on fertility: more reasons to stop smoking. Semin Reprod Med 26(2):204–212

Dain L, Auslander R, Dirnfeld M (2011) The effect of paternal age on assisted reproduction outcome. Fertil Steril 95(1):1–8

De la Rochebrochard E, Thonneau P (2003) Paternal age≥ 40 years: an important risk factor for infertility. Am J Obstet Gynecol 189(4):901–905

de La Rochebrochard E, de Mouzon J, Thépot F, Thonneau P, French National IVF Registry (FIVNAT) Association (2006) Fathers over 40 and increased failure to conceive: the lessons of in vitro fertilization in France. Fertil Steril 85(5):1420–1424

De Mouzon J, Goossens V, Bhattacharya S et al (2010) Assisted reproductive technology in Europe, 2006: results generated from European registers by ESHRE. Hum Reprod 25(8):1851–1862

Depalo R, Garruti G, Totaro I, Panzarino M, Vacca MP, Giorgino F, Selvaggi LE (2011) Oocyte morphological abnormalities in overweight women undergoing in vitro fertilization cycles. Gynecol Endocrinol 27(11):880–884

Dias TR, Alves MG, Bernardino RL, Martins AD, Moreira AC, Silva, J, …, Oliveira PF (2015) Dose-dependent effects of caffeine in human Sertoli cells metabolism and oxidative profile: Relevance for male fertility. Toxicology 328:12–20

Dokras A, Baredziak L, Blaine J, Syrop C, VanVoorhis BJ, Sparks A (2006) Obstetric outcomes after in vitro fertilization in obese and morbidly obese women. Obstet Gynecol 108(1):61–69

Dovey S, Sneeringer RM, Penzias AS (2008) Clomiphene citrate and intrauterine insemination: analysis of more than 4100 cycles. Fertil Steril 90(6):2281–2286

Duits FH, van Wely M, van der Veen F, Gianotten J (2010) Healthy overweight male partners of subfertile couples should not worry about their semen quality. Fertil Steril 94(4):1356–1359

Dumoulin JC, Land JA, Van Montfoort AP, Nelissen EC, Coonen E, Derhaag JG, …, Evers JL (2010a) Effect of in vitro culture of human embryos on birthweight of newborns. Hum Reprod 25(3):605–612

Dunson DB, Baird DD, Colombo B (2004) Increased infertility with age in men and women. Obstet Gynecol 103(1):51–56

Erel CT, Senturk LM (2009) The impact of body mass index on assisted reproduction. Curr Opin Obstet Gynecol 21(3):228–235

ESHRE Task Force on Ethics and Law (2010) Lifestyle-related factors and access to medically assisted reproduction. Hum Reprod 25:578–583

Eskild A, Monkerud L, Tanbo T (2013) Birthweight and placental weight; do changes in culture media used for IVF matter? Comparisons with spontaneous pregnancies in the corresponding time periods. Hum Reprod 28(12):3207–3214

Fauque P, Jouannet P, Lesaffre C, Ripoche MA, Dandolo L, Vaiman D, Jammes H (2007) Assisted Reproductive Technology affects developmental kinetics, H19 Imprinting Control Region methylation and H19 gene expression in individual mouse embryos. BMC Dev Biol 7(1):1

Ford WCL, North K, Taylor H, Farrow A, Hull MGR, Golding J (2000) Increasing paternal age is associated with delayed conception in a large population of fertile couples: evidence for declining fecundity in older men. Hum Reprod 15(8):1703–1708

Frattarelli JL, Miller KA, Miller BT, Elkind-Hirsch K, Scott RT (2008) Male age negatively impacts embryo development and reproductive outcome in donor oocyte assisted reproductive technology cycles. Fertil Steril 90(1):97–103

Fulka H, Mrazek M, Tepla O, Fulka J (2004) DNA methylation pattern in human zygotes and developing embryos. Reproduction 128(6):703–708

Fulka H, St John JC, Fulka J, Hozák P (2008) Chromatin in early mammalian embryos: achieving the pluripotent state. Differentiation 76(1):3–14

Gaur DS, Talekar MS, Pathak VP (2010) Alcohol intake and cigarette smoking: impact of two major lifestyle factors on male fertility. Indian J Pathol Microbiol 53(1):35

Gissler M, Silverio MM, Hemminki E (1995) In-vitro fertilization pregnancies and perinatal health in Finland 1991–1993. Hum Reprod 10(7):1856–1861

Goldberg AD, Allis CD, Bernstein E (2007) Epigenetics: a landscape takes shape. Cell 128(4): 635–638

Griffin J, Emery BR, Huang I, Peterson CM, Carrell DT (2006) Comparative analysis of follicle morphology and oocyte diameter in four mammalian species (mouse, hamster, pig, and human). J Exp & Clin Assist Reprod 3(1):2

Grindler NM, Moley KH (2013) Maternal obesity, infertility and mitochondrial dysfunction: potential mechanisms emerging from mouse model systems. Mol Hum Reprod 19(8):486–494

Gunby J, Bissonnette F, Librach C, Cowan L, Canadian IDG (2010) Assisted reproductive technologies (ART) in Canada: 2006 results from the Canadian ART register. Fertil Steril 93(7):2189–2201

Gunby J, Bissonnette F, Librach C, Cowan L, Canadian, I. D. G (2011) Assisted reproductive technologies (ART) in Canada: 2007 results from the Canadian ART Register. Fertil Steril 95(2):542–547

Guo L, Choufani S, Ferreira J, Smith A, Chitayat D, Shuman C, …, Weksberg R (2008) Altered gene expression and methylation of the human chromosome 11 imprinted region in small for gestational age (SGA) placentae. Dev Biol 320(1):79–91

Hackshaw A, Rodeck C, Boniface S (2011) Maternal smoking in pregnancy and birth defects: a systematic review based on 173 687 malformed cases and 11.7 million controls. Hum Reprod Update 17(5):589–604

Hammiche F, Laven JS, van Mil N, de Cock M, de Vries JH, Lindemans J, …, Steegers-Theunissen RP (2011) Tailored preconceptional dietary and lifestyle counselling in a tertiary outpatient clinic in the Netherlands. Hum Reprod 9:2432–2441.

Hammoud AO, Gibson M, Peterson CM, Meikle AW, Carrell DT (2008a) Impact of male obesity on infertility: a critical review of the current literature. Fertil Steril 90(4):897–904

Hansen KR, Knowlton NS, Thyer AC, Charleston JS, Soules MR, Klein NA (2008) A new model of reproductive aging: the decline in ovarian non-growing follicle number from birth to menopause. Hum Reprod 23:699–708

Hansen M, Kurinczuk JJ, Milne E, de Klerk N, Bower C (2013) Assisted reproductive technology and birth defects: a systematic review and meta-analysis. Hum Reprod Update 19:330–353

Harris ID, Missmer SA, Hornstein MD (2010) Poor success of gonadotropin-induced controlled ovarian hyperstimulation and intrauterine insemination for older women. Fertil Steril 94(1):144–148

Hart R, Norman RJ (2013a) The longer-term health outcomes for children born as a result of IVF treatment: Part I–General health outcomes. Hum Reprod Update 19(3):232–243

Hart R, Norman RJ (2013b) The longer-term health outcomes for children born as a result of IVF treatment. Part II–Mental health and development outcomes. Hum Reprod Update 19(3):244–250

Hassan MA, Killick SR (2003) Effect of male age on fertility: evidence for the decline in male fertility with increasing age. Fertil Steril 79:1520–1527

Heisey AS, Bell EM, Herdt-Losavio ML, Druschel C (2015) Surveillance of congenital malformations in infants conceived through assisted reproductive technology or other fertility treatments. Birth Defects Res Part A: Clin Mol Teratol 103(2):119–126

Helmerhorst FM, Perquin DA, Donker D, Keirse MJ (2004) Perinatal outcome of singletons and twins after assisted conception: a systematic review of controlled studies. BMJ 328 (7434):261

Hodgetts VA, Morris RK, Francis A, Gardosi J, Ismail KM (2015) Effectiveness of folic acid supplementation in pregnancy on reducing the risk of small-for-gestational age neonates: a population study, systematic review and meta-analysis. BJOG Int J Obstet Gynaecol 122(4):478–490

Homan GF, Davies M et al (2007) The impact of lifestyle factors on reproductive performance in the general population and those undergoing infertility treatment: a review. Hum Reprod Update 13(3):209–223

Hossein-nezhad A, Holick MF (2012) Optimize dietary intake of vitamin D: an epigenetic perspective. Curr Opin Clin Nutr & Metab Care 15(6):567–579

Hourvitz A, Machtinger R, Maman E, Baum M, Dor J, Levron J (2009) Assisted reproduction in women over 40years of age: how old is too old? Reprod BioMed Online 19(4):599–603

Huestis MA, Choo RE (2002) Drug abuse's smallest victims: in utero drug exposure. Forensic Sci Int 128(1):20–30

Hull MG, North K et al (2000) Delayed conception and active and passive smoking. Fertil Steril 74:725–733

Hutt KJ, Albertini DF (2007) An oocentric view of folliculogenesis and embryogenesis. Reprod BioMed Online 14(6):758–764

Hutt KJ, Shi Z, Albertini DF, Petroff BK (2008) The environmental toxicant 2, 3, 7, 8-tetrachlorodibenzo-p-dioxin disrupts morphogenesis of the rat pre-implantation embryo. BMC Dev Biol 8(1):1

Ishii T (2014) Potential impact of human mitochondrial replacement on global policy regarding germline gene modification. Reprod BioMed Online 29(2):150–155

Iwahata H, Hashimoto S, Inoue M, Inoue T, Ito K, Nakaoka Y et al (2015) Neonatal outcomes after the implantation of human embryos vitrified using a closed-system device. J Assist Reprod Genet 32(4):521–526

Jackson RA, Gibson KA, Wu YW, Croughan MS (2004) Perinatal outcomes in singletons following in vitro fertilization: a meta-analysis. ObstetGynecol 103(3):551–563

Jafarabadi M (2007) Episodic air pollution is associated with increased DNA fragmentation in human sperm without other changes in semen quality. Hum Reprod 22(12):3263–3263

Jensen TK, Swan S, Jørgensen N, Toppari J, Redmon B et al (2014) Alcohol and male reproductive health: a cross-sectional study of 8344 healthy men from Europe and the USA. Hum Reprod 29:1801–1809

Jensen TK, Swan SH, Skakkebæk NE, Rasmussen S, Jørgensen N (2010) Caffeine intake and semen quality in a population of 2,554 young Danish men. Am J Epidemiol 171(8):883–891

Johnson NP, Bagrie EM, Coomarasamy A, Bhattacharya S, Shelling AN, Jessop S, …, Khan KS (2006) Ovarian reserve tests for predicting fertility outcomes for assisted reproductive technology: the International Systematic Collaboration of Ovarian Reserve Evaluation protocol for a systematic review of ovarian reserve test accuracy. BJOG Int J Obstet Gynaecol 113(12):1472–1480

Joubert BR, Herman T, Felix JF, Bohlin J, Ligthart S, Beckett E, Tiemeier H, Joyce BM, Uitterlinden AG, Hofman A, Håberg SE (2016) Maternal plasma folate impacts differential DNA methylation in an epigenome-wide meta-analysis of newborns. Nat Commun 7. DOI:10.1038/ncomms10577

Juárez SP, Merlo J (2013) Revisiting the effect of maternal smoking during pregnancy on offspring birthweight: a quasi-experimental sibling analysis in Sweden. PLoS One 8(4):e61734

Katari S, Turan N, Bibikova M, Erinle O, Chalian R, Foster M, …, Sapienza C (2009) DNA methylation and gene expression differences in children conceived in vitro or in vivo. Hum Mol Genet 18(20):3769–3778

Kazaura MR, Lie RT (2002) Down's syndrome and paternal age in Norway. Paediatr Perinat Epidemiol 16(4):314–319

Khosla S, Dean W, Reik W, Feil R (2001a) Culture of preimplantation embryos and its long-term effects on gene expression and phenotype. Hum Reprod Update 7(4):419–427

Khosla S, Dean W, Brown D, Reik W, Feil R (2001b) Culture of preimplantation mouse embryos affects fetal development and the expression of imprinted genes. BiolReprod 64(3):918–926

Kidd SA, Eskenazi B, Wyrobek AJ (2001) Effects of male age on semen quality and fertility: a review of the literature. Fertil Steril 75(2):237–248

Kissin DM, Zhang Y, Boulet SL, Fountain C, Bearman P, Schieve L, …, Jamieson DJ (2015) Association of assisted reproductive technology (ART) treatment and parental infertility diagnosis with autism in ART-conceived children. Hum Reprod 30(2):454–465

Kleinhaus K, Perrin M, Friedlander Y, Paltiel O, Malaspina D, Harlap S (2006) Paternal age and spontaneous abortion. Obstet Gynecol 108(2):369–377

Klemetti R, Gissler M, Sevón T, Koivurova S, Ritvanen A, Hemminki E (2005) Children born after assisted fertilization have an increased rate of major congenital anomalies. Fertil Steril 84(5):1300–1307

Klonoff-Cohen HS, Natarajan L (2004) The effect of advancing paternal age on pregnancy and live birth rates in couples undergoing in vitro fertilization or gamete intrafallopian transfer. Am J Obstet Gynecol 191(2):507–514

Klonoff-Cohen H, Bleha J, Lam-Kruglick P (2002) A prospective study of the effects of female and male caffeine consumption on the reproductive endpoints of IVF and gamete intra-Fallopian transfer. Hum Reprod 17(7):1746–1754

Klonoff-Cohen H, Lam-Kruglick P, Gonzalez C (2003) Effects of maternal and paternal alcohol consumption on the success rates of in vitro fertilization and gamete intrafallopian transfer. Fertil Steril 79(2):330–339

Kobayashi H, Hiura H, John RM, Sato A, Otsu E, Kobayashi N, …, Yaegashi N (2009) DNA methylation errors at imprinted loci after assisted conception originate in the parental sperm. Eur J Hum Genet 17(12):1582–1591

Koopman WJ, Willems PH, Smeitink JA (2012) Monogenic mitochondrial disorders. N Engl J Med 366(12):1132–1141

Kovac JR, Addai J, Smith RP, Coward RM, Lamb DJ, Lipshultz LI (2013) The effects of advanced paternal age on fertility. Asian J Androl 15(6):723–728

Kovatsi L, Fragou D, Samanidou V, Njau S, Kouidou S (2011) Drugs of abuse: epigenetic mechanisms in toxicity and addiction. Curr Med Chem 18(12):1765–1774

Leridon H (2004) Can assisted reproduction technology compensate for the natural decline in fertility with age? A model assessment. Hum Reprod 19(7):1548–1553

Li Z, Ye R, Zhang L, Li H, Liu J, Ren A (2014) Periconceptional folic acid supplementation and the risk of preterm births in China: a large prospective cohort study. Int J Epidemiol 43(4):1132–1139

Ludwig M (2009) Are adverse outcomes associated with assisted reproduction related to the technology or couples' subfertility? NatClinPractUrol 6(1):8–9

Ludwig M, Katalinic A, Gross S, Sutcliffe A, Varon R, Horsthemke B (2005) Increased prevalence of imprinting defects in patients with Angelman syndrome born to subfertile couples. J Med Genet 42(4):289–291

Luke B, Brown MB, Stern JE, Missmer SA, Fujimoto VY, Leach R (2011) Female obesity adversely affects assisted reproductive technology (ART) pregnancy and live birth rates. Hum Reprod 26(1):245–252

Luna M, Finkler E, Barritt J, Bar-Chama N, Sandler B, Copperman AB, Grunfeld L (2009) Paternal age and assisted reproductive technology outcome in ovum recipients. Fertil Steril 92(5):1772–1775

MacDonald AA, Herbison GP, Showell M, Farquhar CM (2010) The impact of body mass index on semen parameters and reproductive hormones in human males: a systematic review with meta-analysis. Hum Reprod Update 16(3):293–311

Maconochie N, Doyle P, Prior S, Simmons R (2007) Risk factors for first trimester miscarriage—results from a UK-population-based case–control study. BJOG Int J Obstet Gynaecol 114(2):170–186

Maher ER (2005) Imprinting and assisted reproductive technology. Hum Mol Genet 14(suppl 1):R133–R138

Maheshwari A, Stofberg L, Bhattacharya S (2007) Effect of overweight and obesity on assisted reproductive technology—a systematic review. Hum Reprod Update 13(5):433–444

Mahsoudi B, Li A, O'Neill C (2007) Assessment of the long-term and transgenerational consequences of perturbing preimplantation embryo development in mice. Biol Reprod 77(5):889–896

Mäkinen S, Söderström-Anttila V, Vainio J, Suikkari AM, Tuuri T (2013) Does long in vitro culture promote large for gestational age babies? Hum Reprod 28(3):828–834

Marczylo EL, Amoako AA, Konje JC, Gant TW, Marczylo TH (2012) Smoking induces differential miRNA expression in human spermatozoa: a potential transgenerational epigenetic concern? Epigenetics 7(5):432–439

Marques CJ, Carvalho F, Sousa M, Barros A (2004) Genomic imprinting in disruptive spermatogenesis. Lancet 363(9422):1700–1702

McCowan L, Horgan RP (2009) Risk factors for small for gestational age infants. Best Pract & Res Clin Obstet & Gynaecol 23(6):779–793

McDonald SD, Han Z, Mulla S, Murphy KE, Beyene J, Ohlsson A (2009) Preterm birth and low birth weight among in vitro fertilization singletons: a systematic review and meta-analyses. Eur J Obstet Gynecol Reprod Biol 146(2):138–148

Meeker JD, Benedict MD (2013) Infertility, pregnancy loss and adverse birth outcomes in relation to maternal secondhand tobacco smoke exposure. Curr women's health reviews 9(1):41

Messerlian C, Maclagan L, Basso O (2012) Infertility and the risk of adverse pregnancy outcomes: a systematic review and meta-analysis. Hum Reprod 28:125–137

Miller WH, Hyatt MC (1992) Perinatal substance abuse. Am J Drug Alcohol Abuse 18(3):247–261

Morgan HD, Jin XL, Li A, Whitelaw E, O'Neill C (2008) The culture of zygotes to the blastocyst stage changes the postnatal expression of an epigenetically labile allele, agouti viable yellow, in mice. BiolReprod 79(4):618–623

MRC Vitamin Study Research Group (1991) Prevention of neural tube defects: results of the Medical Research Council Vitamin Study. Lancet 338(8760):131–137

Mtango NR, Potireddy S, Latham KE (2009) Expression of microRNA processing machinery genes in rhesus monkey oocytes and embryos of different developmental potentials. Mol Reprod Dev 76(3):255–269

Muthusami KR, Chinnaswamy P (2005) Effect of chronic alcoholism on male fertility hormones and semen quality. Fertil Steril 84(4):919–924

Myklestad K, Vatten LJ, Magnussen EB, Salvesen KÅ, Smith GD, Romundstad PR (2012) Offspring birth weight and cardiovascular risk in parents—a population-based HUNT 2 Study. Am J Epidemiol 175(6):546–555

Nafee TM, Farrell WE, Carroll WD, Fryer AA, Ismail KMK (2008) Review article: epigenetic control of fetal gene expression. BJOG Int J Obstet Gynaecol 115(2):158–168

Nagaoka SI, Hassold TJ, Hunt PA (2012) Human aneuploidy: mechanisms and new insights into an age-old problem. Nat Rev Genet 13(7):493–504

Nelissen EC, Van Montfoort AP, Coonen E, Derhaag JG, Geraedts JP, Smits LJ, Land JA, Evers JL, Dumoulin JC (2012) Further evidence that culture media affect perinatal outcome: findings after transfer of fresh and cryopreserved embryos. Hum Reprod 27:1966–1976

Neri M, Bello S, Turillazzi E, Riezzo I (2015) Drugs of abuse in pregnancy, poor neonatal development, and future neurodegeneration. Is oxidative stress the culprit? Curr Pharm Des 21(11):1358–1368

Nguyen RH, Wilcox AJ, Skjærven R, Baird DD (2007) Men's body mass index and infertility. Hum Reprod 22(9):2488–2493
Olson CK, Keppler-Noreuil KM, Romitti PA, Budelier WT, Ryan G, Sparks AE, Van Voorhis BJ (2005) In vitro fertilization is associated with an increase in major birth defects. Fertil Steril 84(5):1308–1315
Otero C, Boles S, Young NK, Dennis K (2004) Methamphetamine: addiction, treatment, outcomes and implications. US Department of Health and Human Services. SAMHSA, Rockville, MD
Ozturk S, Yaba-Ucar A, Sozen B, Mutlu D, Demir N (2016) Superovulation alters embryonic poly (A)-binding protein (Epab) and poly (A)-binding protein, cytoplasmic 1 (Pabpc1) gene expression in mouse oocytes and early embryos. Reprod Fertil Dev 28(3):375–383
Paasch U, Grunewald S, Kratzsch J, Glander HJ (2010) Obesity and age affect male fertility potential. Fertil Steril 94(7):2898–2901
Pasquali R (2006) Obesity and androgens: facts and perspectives. Fertil Steril 85(5):1319–1340
Pasquali R, Patton L, Gambineri A (2007) Obesity and infertility. Curr Opin Endocrinol, Diabetes Obes 14(6):482–487
Pasqualotto FF, Umezu FM, Salvador M, Borges E, Sobreiro BP, Pasqualotto EB (2008) Effect of cigarette smoking on antioxidant levels and presence of leukocytospermia in infertile men: a prospective study. Fertil Steril 90(2):278–283
Paulson RJ, Milligan RC, Sokol RZ (2001) The lack of influence of age on male fertility. Am J Obstet Gynecol 184(5):818–824
Pellestor F, Andréo B, Arnal F, Humeau C, Demaille J (2003) Maternal aging and chromosomal abnormalities: new data drawn from in vitro unfertilized human oocytes. Hum Genet 112(2):195–203
Phibbs CS, Bateman DA, Schwartz RM (1991) The neonatal costs of maternal cocaine use. JAMA 266(11):1521–1526
Practice Committee of the American Society for Reproductive Medicine (2008) Obesity and reproduction: an educational bulletin. Fertil Steril 90(5):S21–S29
Qiao J, Chen Y, Yan LY, Yan J, Liu P, Sun QY (2010) Changes in histone methylation during human oocyte maturation and IVF-or ICSI-derived embryo development. Fertil Steril 93(5):1628–1636
Qin DD, Yuan W, Zhou WJ, Cui YQ, Wu JQ, Gao ES (2007) Do reproductive hormones explain the association between body mass index and semen quality? Asian J Androl 9(6):827–834
Rahnama F, Shafiei F, Gluckman PD, Mitchell MD, Lobie PE (2006) Epigenetic regulation of human trophoblastic cell migration and invasion. Endocrinology 147(11):5275–5283
Ramlau-Hansen CH, Thulstrup AM, Aggerholm AS, Jensen MS, Toft G, Bonde JP (2007a) Is smoking a risk factor for decreased semen quality? A cross-sectional analysis. Hum Reprod 22(1):188–196
Ramlau-Hansen CH, Thulstrup AM, Nohr EA, Bonde JP, Sørensen TIA, Olsen J (2007b) Subfecundity in overweight and obese couples. Hum Reprod 22(6):1634–1637
Reik W, Walter J (2001) Genomic imprinting: parental influence on the genome. Nat Rev Genet 2(1):21–32
Reinhardt K, Dowling DK, Morrow EH (2013) Mitochondrial replacement, evolution, and the clinic. Science 341(6152):1345–1346
Romundstad LB, Romundstad PR, Sunde A, von Düring V, Skjærven R, Vatten LJ (2006) Increased risk of placenta previa in pregnancies following IVF/ICSI; a comparison of ART and non-ART pregnancies in the same mother. Hum Reprod 21(9):2353–2358
Romundstad LB, Romundstad PR, Sunde A, von Düring V, Skjærven R, Gunnell D, Vatten LJ (2008) Effects of technology or maternal factors on perinatal outcome after assisted fertilisation: a population-based cohort study. Lancet 372(9640):737–743
Rubio C, Simon C, Vidal F, Rodrigo L, Pehlivan T, Remohi J, Pellicer A (2003) Chromosomal abnormalities and embryo development in recurrent miscarriage couples. Hum Reprod 18(1): 182–188
Said TM, Ranga G, Agarwal A (2005) Relationship between semen quality and tobacco chewing in men undergoing infertility evaluation. Fertil Steril 84(3):649–653

Sallmén M, Sandler DP, Hoppin JA, Blair A, Baird DD (2006) Reduced fertility among overweight and obese men. Epidemiology 17(5):520–523

SAMHSA (2015) Behavioral Health Trends in the United States: Results from the 2014 National Survey on Drug Use and Health. https://www.samhsa.gov/data/sites/default/files/NSDUH-FRR1-2014/NSDUH-FRR1-2014.pdf

Santos F, Dean W (2004) Epigenetic reprogramming during early development in mammals. Reproduction 127(6):643–651

Santos F, Dean W (2006) Using immunofluorescence to observe methylation changes in mammalian preimplantation embryos. Nucl Reprogramming: Methods and Protocols 325:129–138

Santos MA, Kuijk EW, Macklon NS (2010) The impact of ovarian stimulation for IVF on the developing embryo. Reproduction 139(1):23–34

Sato A, Otsu E, Negishi H, Utsunomiya T, Arima T (2007) Aberrant DNA methylation of imprinted loci in superovulated oocytes. Hum Reprod 22(1):26–35

Schaefer AM, Taylor RW, Turnbull DM, Chinnery PF (2004) The epidemiology of mitochondrial disorders—past, present and future. Biochimica et Biophysica Acta (BBA)-Bioenergetics 1659(2):115–120

Schier AF (2007) The maternal-zygotic transition: death and birth of RNAs. Science 316(5823): 406–407

Schieve LA, Meikle SF, Ferre C, Peterson HB, Jeng G, Wilcox LS (2002) Low and very low birth weight in infants conceived with use of assisted reproductive technology. N Engl J Med 2002(346):731–737

Schieve LA, Cohen B, Nannini A, Ferre C, Reynolds MA, Zhang Z, …, Massachusetts Consortium for Assisted Reproductive Technology Epidemiologic Research (MCARTER) (2007) A population-based study of maternal and perinatal outcomes associated with assisted reproductive technology in Massachusetts. Matern Child Health J 11(6):517–525

Schieve LA, Rasmussen SA, Buck GM, Schendel DE, Reynolds MA, Wright VC (2004) Are children born after assisted reproductive technology at increased risk for adverse health outcomes? Obstet Gynecol 103(6):1154–1163

Seggers J, de Walle HE, Bergman JE, Groen H, Hadders-Algra M, Bos ME, …, Haadsma ML (2015) Congenital anomalies in offspring of subfertile couples: a registry-based study in the northern Netherlands. Fertil Steril 103(4):1001–1010

Sermondade N, Elloumi H, Berthaut I, Mathieu E, Delarouzière V, Ravel C, Mandelbaum J (2010) Progressive alcohol-induced sperm alterations leading to spermatogenic arrest, which was reversed after alcohol withdrawal. Reprod BioMed Online 20(3):324–327

Sermondade N, Faure C, Fezeu L, Lévy R, Czernichow S (2012) Obesity and increased risk for oligozoospermia and azoospermia. Arch Intern Med 172(5):440–442

Sermondade N, Faure C, Fezeu L, Shayeb AG, Bonde JP, Jensen TK, …, Chavarro JE (2013) BMI in relation to sperm count: an updated systematic review and collaborative meta-analysis. Hum Reprod Update 19(3):221–231

Shi L, Wu J (2009) Epigenetic regulation in mammalian preimplantation embryo development. ReprodBiolEndocrinol 7:59

Soares SR et al (2007) Cigarette smoking affects uterine receptiveness. Hum Reprod 22(2):543–547

Sobreiro BP, Lucon AM, Pasqualotto FF, Hallak J, Athayde KS et al (2005) Semen analysis in fertile patients undergoing vasectomy: reference values and variations according to age, length of sexual abstinence, seasonality, smoking habits and caffeine intake. Sao Paulo Med J 123:161–166

Sofikitis N et al (2000) Effects of nicotine on sperm motility, membrane function and fertilizing capacity in vitro. Urol Res 28(6):370–375

Soubry A, Schildkraut JM, Murtha A, Wang F, Huang Z, Bernal A et al (2013) Paternal obesity is associated with IGF2 hypomethylation in newborns: results from a Newborn Epigenetics Study (NEST) cohort. BMC Med 11:29

Steegers-Theunissen RPM, Twigt J, Pestinger V, Sinclair KD (2013) The periconceptional period, reproduction and long-term health of offspring: the importance of one-carbon metabolism. Hum Reprod Update 19:640–655

Stewart AF, Kim ED (2001) Fertility concerns for the aging male. Urology 78:496–499
Strachan T, Read A (2003) Human molecular genetics, 3rd edn. Garland Science, Taylor and Francis Group, London
Surén P, Roth C, Bresnahan M, Haugen M, Hornig M, Hirtz D, …, Schjølberg S (2013) Association between maternal use of folic acid supplements and risk of autism spectrum disorders in children. JAMA 309(6):570–577
Sutcliffe AG, Ludwig M (2007) Outcome of assisted reproduction. Lancet 370(9584):351–359
Sutcliffe AG, Peters CJ, Bowdin S, Temple K, Reardon W, Wilson L, …, Maher ER (2006) Assisted reproductive therapies and imprinting disorders—a preliminary British survey. Hum Reprod 21(4):1009–1011
Tanbo T, Dale PO, Lunde O, Moe N, ÅRBYHOLM T (1995) Obstetric 0 singleton pregnancies after assisted reproduction. Obstet Gynecol 86(2):188–192
Tang F, Kaneda M, O'Carroll D, Hajkova P, Barton SC, Sun YA, …, Surani MA (2007) Maternal microRNAs are essential for mouse zygotic development. Genes Dev 21(6):644–648
Vergouw CG, Kostelijk EH, Doejaaren E, Hompes PG, Lambalk CB, Schats R (2012) The influence of the type of embryo culture medium on neonatal birthweight after single embryo transfer in IVF. Hum Reprod 27:2619–2626
Vermeiden JP, Bernardus RE (2013) Are imprinting disorders more prevalent after human in vitro fertilization or intracytoplasmic sperm injection? Fertil Steril 99(3):642–651
Vogel G (1997) Cocaine wreaks subtle damage on developing brains. Science 278(5335):38–39
Warburton D (2005) Biological aging and the etiology of aneuploidy. Cytogenet Genome Res 111(3–4):266–272
Watkins AJ, Platt D, Papenbrock T, Wilkins A, Eckert JJ, Kwong WY, …, Fleming TP (2007) Mouse embryo culture induces changes in postnatal phenotype including raised systolic blood pressure. Proc Natl Acad Sci 104(13):5449–5454
Watkins A, Papenbrock T, Fleming TP (2008) The preimplantation embryo: handle with care. Semin Reprod Med 26(2):175–185
Waylen AL, Metwally M, Jones GL, Wilkinson AJ, Ledger WL (2009) Effects of cigarette smoking upon clinical outcomes of assisted reproduction: a meta-analysis. Hum Reprod Update 15(1):31–44
Wesselink AK, Wise LA, Rothman KJ, Hahn KA, Mikkelsen EM, Mahalingaiah S, Hatch EE (2016) Caffeine and caffeinated beverage consumption and fecundability in a preconception cohort. Reprod Toxicol 62:39–45
Wijers CH, van Rooij IA, Rassouli R, Wijnen MH, Broens PM, Sloots CE, …, Roeleveld N (2015) Parental Subfertility, Fertility Treatment, and the Risk of Congenital Anorectal Malformations. Epidemiology 26(2):169–176
Winter E, Wang J, Davies MJ, Norman R (2002) Early pregnancy loss following assisted reproductive technology treatment. Hum Reprod 17(12):3220–3223
Wisborg K, Ingerslev HJ, Henriksen TB (2010) IVF and stillbirth: a prospective follow-up study. Hum Reprod 25(5):1312–1316
Yang Y, Bai W, Zhang L, Yin G, Wang X, Wang J, …, Yao YQ (2008) Determination of microRNAs in mouse preimplantation embryos by microarray. Dev Dyn 237(9):2315–2327
Yang X, Chen H, Du Y, Wang S, Wang Z (2016) Periconceptional folic acid fortification for the risk of gestational hypertension and pre-eclampsia: a meta-analysis of prospective studies. Matern Child Nutr 12(4):669–679
Zegers-Hochschild F, Adamson GD, de Mouzon J, Ishihara O, Mansour R, Nygren K, …, Van der Poel S (2009) The international committee for monitoring assisted reproductive technology (ICMART) and the world health organization (WHO) revised glossary on ART terminology, 2009. Hum Reprod 24, (11):2683–2687
Zhu JL, Basso O, Obel C, Bille C, Olsen J (2006) Infertility, infertility treatment, and congenital malformations: Danish national birth cohort. BMJ 333(7570):679
Zuckerman B, Frank DA, Hingson R, Amaro H, Levenson SM, Kayne H, …, Cabral H (1989) Effects of maternal marijuana and cocaine use on fetal growth. N Engl J Med 320(12):762–768

The Importance of the Periconception Period: Immediate Effects in Cattle Breeding and in Assisted Reproduction Such as Artificial Insemination and Embryo Transfer

Mieke Van Eetvelde, Sonia Heras, J.L.M.R. Leroy, Ann Van Soom, and Geert Opsomer

Abstract In livestock breeding, the successful outcome is largely depending on the "periconception environment" which, in a narrow sense, refers to the genital tract, where gametogenesis and embryogenesis occur. During these early stages of development, gametes and embryos are known to be particularly sensitive to alterations in their microenvironment. However, as the microenvironment somehow reflects what is going on in the external world, we must widen our definition of "periconception environment" and refer to all events taking place around the time of conception, including metabolic state and health and nutrition of the dam. In modern dairy cows that have to manage an optimal reproductive performance with continued growth and high milk yield, the periconception period is particularly challenging. The metabolic priority for growth and lactation is known to generate adverse conditions hampering optimal ovarian function, oocyte maturation, and development of embryo/fetus. In addition, by using artificial reproductive technologies (ARTs), gametes and/or embryos of livestock are exposed to unnatural conditions outside the male and female genital tract. Artificial insemination, the most widely used technique, is currently yielding pregnancy rates similar to natural mating, and calves

---

Focus: Discussing the immediate effects that the periconception environment can have on conception success, the rate of fertility, and the establishment of pregnancy in livestock.

M. Van Eetvelde and S. Heras contributed equally to this work.

M. Van Eetvelde • S. Heras • A. Van Soom (✉) • G. Opsomer
Department of Reproduction, Obstetrics and Herd Health, Faculty of Veterinary Medicine, Ghent University, Salisburylaan 133, B-9820 Merelbeke, Belgium
e-mail: mieke.vaneetvelde@ugent.be; Sonia.HerasGarcia@UGent.be; ann.vansoom@ugent.be; geert.opsomer@ugent.be

J.L.M.R. Leroy
University of Antwerp, Campus Drie Eiken,
Universiteitsplein 1 D.U.010, 2610 Wilrijk, Belgium
e-mail: jo.leroy@uantwerpen.be

A. Fazeli, W.V. Holt (eds.), *Periconception in Physiology and Medicine*,
Advances in Experimental Medicine and Biology 1014,
DOI 10.1007/978-3-319-62414-3_3

produced by AI are equally viable after natural mating. In contrast, other ART, such as multiple ovulation and embryo transfer, have been reported to induce changes in gene expression and DNA methylation patterns with potential consequences for development.

Finally, the "periconceptional" environment has been shown to not only influence the successful establishment of pregnancy but also the long-term health and productivity of the offspring. Hence, the optimization of management around the time of conception might open doors to improve animal production and product quality.

Keywords Epigenetics • Artificial insemination • Embryo transfer • Periconception environment • Farm animals

1 Introduction

Livestock breeding is seemingly not very difficult. Unlike in human reproduction, dairy and beef cattle can be selected for fertility traits (Bowley et al. 2015), and animals suffering from suboptimal fertility are eventually removed from the breeding system, unless they are of very high genetic value. However, cattle breeding is nowadays affected by several factors, such as animal management conditions (Shahinfar et al. 2014) and climate change (Burthe et al. 2011; Gauly et al. 2013). These factors can influence or change the "periconception environment." Periconception environment can be interpreted in a broader sense, and all the events that take place during this period are extensively discussed in the various chapters of this book. In the present chapter, we will use the expression "periconception environment" in the narrow sense of the word, first to specifically refer to the male and female genital tract, where gametogenesis and embryogenesis occur. In the second phase, we will also discuss "periconception environment" in a broader sense, when referring to the physiology of the animal and its environment.

At first sight, it may appear that the oocyte, maturing in the ovarian follicle, and the embryo, developing within the oviduct and uterus, are well isolated from issues related to the external environment and therefore obviously benefit from being in a protected environment. However, there is a growing realization that the period during which the oocyte is developing in the ovarian follicle, and also during the first few weeks of embryo development, is one in which gametes and embryos are particularly sensitive to changes occurring in their "protected" environment. Apparently, this "protected environment" somehow reflects what is going on in the external world. Features such as dietary changes; general health issues and inflammation, heat stress, and metabolic stress; and environmental pollution may affect the intrinsic quality of gametes, preimplantation embryos, or even fetal development and thus negatively impact on conception rates and the chance of a successful establishment of pregnancy in livestock. Furthermore, there is currently more and more evidence, both from studies in humans and in a variety of animal species, that factors influencing the periconception environment not only have consequences on the short term, deteriorating pregnancy results and neonatal survival, but may have a distinct effect on the long term by impairing animal health and productivity later in life. Hence, in livestock there is a considerable interest in this concept, also known as developmental programming, since it may open doors to improve animal production and product quality when implemented in general management strategies (Wu et al. 2006).

However, with the advent of assisted reproductive technologies (ARTs), we must further widen our definition of the periconception environment. "Assisted reproductive technologies" refer to four generations of technologies, which all have an impact on the periconception environment in the broader sense: (1) artificial insemination (AI), (2) multiple ovulation and embryo transfer (MOET), (3) in vitro embryo production (IVP), and (4) cloning and transgenesis (Thibier 2005). In this chapter, we will only focus on the first three, since they are routinely used in practice, whereas cloning and transgenesis are not, although this may soon change with the construction of an animal cloning factory in Tianjin, China (ref http://www.theguardian.com/world/2015/nov/24/worlds-largest-animal-cloning-factory-can-save-species-says-chinese-founder).

By using ART, gametes and/or embryos of livestock are being exposed to unnatural conditions outside the male or female genital tract, such as exposure to ambient air, to changes in temperature and pH, to nonphysiological hormone balances, to synthetic culture media, and even to cryopreservation or to breaches in the cellular membrane. In this chapter, we will reiterate how these changes in periconception environment may affect fertility in livestock and even health of the offspring, with a focus on dairy cattle.

2 Gametogenesis, Embryogenesis, Establishment of Pregnancy, and Fetal and Placental Development in Cattle

Female calves are born with hundreds of thousands of oocytes (Erickson 1966, cited by Britt 2008), which remain in the ovaries as primordial and primary follicles until puberty. The moment of puberty is significantly affected by breed and body growth rates. Furthermore, a cow is a year-round breeder. When heifers start to cycle at about 12 months of age, about every 3 weeks, a single mature oocyte, surrounded by some expanded cumulus cells, is released into the oviduct. The estimated time course of folliculogenesis in the bovine is about 80–100 days (Britt 2008), from the stage of the primary follicle until the mature Graafian follicle. During this period of oogenesis, follicular waves emerge, but only one dominant follicle is selected to ovulate. Dairy heifers are usually bred at the age of 15 months, just after ovulation.

Bull calves start to produce semen at puberty (9–10 months) and are sexually mature at the age of 2–4 years (Rawlings et al. 2008). In each ejaculate, they produce billions of spermatozoa, and the estimated time course of spermatogenesis is 62 days. Conception rates are high in beef cattle mated by a bull (up to 80%) but significantly lower (40%) in modern high-producing dairy cattle that are currently traditionally bred by AI. In both cases, fertilization of the oocyte occurs within the oviduct. The bovine embryo enters the uterus at about 4–5 days after ovulation, at the early morula stage. During the period preceding embryo attachment, the embryo is free floating and is dependent upon endometrial secretions into the uterine lumen, termed histotroph, both for energy and proteins. The critical period of maternal recognition of pregnancy occurs between days 15 and 18 after ovulation, followed by the initial stages of implantation and early placentation (starting gradually after day 19 (Lonergan and Forde 2015)). In the cow, the placenta attaches to discrete sites of the uterine wall called caruncles. These caruncles are arranged in two dorsal and two ventral rows throughout the length of the uterine horns. The placental membranes attach at these sites via chorionic villi in specific areas called cotyledons. The caruncular-cotyledonary unit is called a placentome and is the functional area of exchanges between the cow and calf. In association with the formation of the placentome, the caruncular area is progressively vascularized to meet the increasing demands of the conceptus. Cattle have been shown to have the synepitheliochorial type of placentation (Wooding and Burton 2008). Fetal survival is dependent upon proper placental growth and vascularity early in pregnancy, while further intrauterine growth is mainly dependent on the placental supply of maternal nutrients and oxygen to the fetus. Therefore, establishment of a functional fetal/placental vascular system is one of the earliest requirements during conceptus development (Vonnahme et al. 2008). However, during pregnancy, the placenta is exposed to a variety of environmental insults which can alter fetal organogenesis and growth, leading to improper pre- and postnatal growth and eventually lower life performance.

Calving rates are typically high in beef cattle after natural mating (80%), and in dairy cattle after artificial insemination (65%) and embryo transfer of in vivo derived embryo (60%), but decrease toward less than 50% after transferring in vitro produced bovine embryos. Interestingly, in the high-producing dairy cow, calving rates after AI have decreased over the last decades to about 40%. In the next paragraphs, we will discuss how these differences can be explained, with an emphasis on dairy cattle.

3 Factors Affecting Periconception Environment Both in Natural Breeding and Assisted Reproduction

3.1 A Dairy Cow Is Not a Typical Cow

To maximize milk production, currently, farmers are stimulated to breed their stock at young age in order to have a first calf at 24 months and subsequently have their cows calved with intervals no longer than 385–400 days. The latter implies dairy cows to be rather atypical since they have to manage the compatibility of optimal reproductive performance and (early) gestation with continued growth or the production of large quantities of milk. To assure a high level of milk production, heifers should be raised to weigh 350–375 kg at 15 months of age, the age at which they should be inseminated in order to allow calving at 24 months (Wathes et al. 2014). A pregnant lactating cow's capacity to care for her embryo is largely determined by the way she partitions nutrients to support embryonic, placental, and fetal development together with her own maintenance and milk production (Wathes 2012). Continued growth, production status, and energy balance are known to have a significant effect on how nutrients are partitioned. Rather than being an absolute shortage of energy substrates per se, this metabolic priority for growth and lactation (after calving) is known to generate adverse conditions hampering optimal ovarian functioning, follicular growth, oocyte maturation, and early embryo development (Leroy et al. 2008a, b). In the next paragraph, we will further focus on "reproductive success" in dairy cattle, which can be defined as the ability to produce a healthy offspring in a timely way.

3.2 Physiological Factors Affecting Reproductive Success in Dairy Cattle

3.2.1 Continued Growth in Adolescent Animals

In general, reproductive capacity of nulliparous heifers is higher when compared to that of multiparous animals. While this finding can be attributed to the fact that oocytes and embryos of the nulliparous heifers have not been challenged by the metabolic stress of milk production, one should also not forget the decisive role of the uterus in terms of pregnancy success. Uteri of nulliparous heifers have not been confronted yet with a parturition event which is, in the vast majority of cases, associated with bacterial infection. Besides the better reproductive performance of nulliparous heifers, significant differences have been noted in terms of production, reproductive capacity, longevity, and resilience against metabolic challenges, between offspring of first and higher parity animals (Banos et al. 2007; González-Recio et al. 2012), with, in most cases, the offspring of first parity animals being in a more favorable condition. The latter results may be interpreted as an indication of the deleterious effects of lactation during conception and early pregnancy since in contrast to multiparous dairy cows, first-parity animals do not lactate.

All too often, however, researchers have used first-parity heifers as non-lactating and hence "negative" controls when examining the effect of lactation and its concomitant metabolic consequences on the animal's reproductive capacity. However, when reproduction has to coincide with continued growth of the first-parity dam, the fetus may face intense competition for nutrients from its mother's own metabolic needs while still growing. Hence, the normal hierarchy of nutrient partitioning between maternal body growth and fetal growth may be altered (Wallace et al. 2006). In sheep, for example, there is a general consensus nowadays that overnutrition during gestation in adolescent ewes gives rise to a lighter progeny, while the dam generally experiences a significant increase in body condition. In this paradigm, rapid maternal growth results in placental growth restriction and often premature delivery of low birth weight lambs when compared with moderately nourished ewes of equivalent age (Wallace et al. 2006).

Since farmers are stimulated to maximize daily growth in their growing young stock in order to maximize milk production in the first and subsequent lactations, they accentuate the mismatch between the milieu for which the offspring is prepared and the milieu the neonates actually experience, which may lead to even more deleterious effects. Examples hereof are well known in human medicine, where it has been shown that babies that had experienced intrauterine growth retardation and thereafter experience a catch-up growth are more prone to reproductive disorders such as polycystic ovarian syndrome (Ibánez et al. 2008). Epidemiological studies both in beef (Funston and Deutscher 2004; Funston et al. 2012) and in dairy cattle (Brickell et al. 2009; Swali and Wathes 2007) have indeed shown that heifers growing fast in the first months of life have a significantly earlier pubarche but need more inseminations to become pregnant, ending up with a similar age at first calving in comparison with their slower growing peers. In this light, we may refer to the "thrifty phenotype hypothesis," which proposes the epidemiological associations between poor fetal and infant growth and the subsequent development of type 2 diabetes and metabolic syndrome. Both of these outcomes result from the effects of poor nutrition in early life, which produces permanent changes in glucose-insulin metabolism (Hales and Barker 2001). This hypothesis may also apply to high-producing dairy cattle.

3.2.2 High Milk Yield and the Concomitant Metabolic Stress in Lactating Animals

The genetic drive to produce large quantities of milk makes modern dairy cows more vulnerable to factors generally known to impair overall health and fertility. Hence, since dairy cows are challenged by such a variety of environmental factors during the period when they should also reproduce, they represent a "natural" model to describe the effects of periconception environmental challenges on their reproductive capacity. Furthermore, the modern dairy cows' reproductive capacity is under extreme pressure especially because of very high rates of (early) embryonic mortality (Wiltbank et al. 2016). The latter might be a reflection of the high number of insults the gametes and early embryos are confronted with during the periconception period (Ribeiro et al. 2016; Leroy et al. 2015).

Modern dairy cows have been predominantly selected for high milk yield in early lactation, which is associated with a very high capacity to mobilize body reserves during this period. Calculations have shown that cows can produce as much as between 120 and 550 kg of milk from body reserves on the basis of energy (average 324 kg). Most cows can cope with this metabolic load, which is defined as "the total energy burden imposed by the synthesis and secretion of milk, which is met by mobilization of body reserves." Metabolic load has, however, been opposed to metabolic stress, which is defined as "the amount of metabolic load that cannot be sustained by body mobilization, leading to the down-regulation of some energetic processes, including those that maintain general health" (Knight et al. 2000). Hence, the "over"-mobilization of body reserves during the period of negative energy balance is a key factor for disease susceptibility in modern dairy cattle. The genetically and hormonally driven body mobilization is further significantly aggravated by the mismatch between the energy need and the cow's capacity to take in energy, the latter often being even further negatively affected by an inadequate adaptation of both the gastrointestinal tract and the overall intermediary metabolism. Maximal feed intake in dairy cows occurs commonly at 6–8 weeks in lactation, which is much later than peak production, causing cows typically to be in negative energy balance for 5–7 weeks postpartum (Tamminga et al. 1997). High milk production per se is not the primary cause to elicit negative effects on health and fertility traits, since the effect mainly seems to depend on the overall farm management and production environment.

Typically, the negative energy balance and concomitant body fat mobilization are characterized by specific alterations in peripheral plasma metabolite concentrations such as high non-esterified fatty acids (NEFAs), low glucose and insulin, and high levels of ketone bodies. In our laboratory, ovum pickup (OPU) has been used to demonstrate that these alterations not only occur in the peripheral circulation but are also reflected in the follicular fluid of the ovaries (Leroy et al. 2004). Applying OPU allowed us to get a better insight into the environment in which both the follicular cells and the oocyte have to mature. The effects of the elevated/lowered concentrations of metabolites associated with high milk yield on follicular cells (Vanholder et al. 2005b, 2006) and oocytes (Leroy et al. 2005, 2006) were subsequently evaluated in the laboratory. Main conclusions were that saturated NEFAs at concentrations found in vivo were able to impair proliferation of granulosa and theca cells (Vanholder et al. 2005b, 2006), while oocytes that had to mature in media in which elevated levels of saturated NEFAs were added were associated with lower fertilization and blastocyst rates (Leroy et al. 2005). Furthermore, when oocytes were matured in vitro in media mimicking ketotic situations, low glucose rather than high levels of ketone bodies seemed to be detrimental for subsequent fertilization and blastocyst formation (Leroy et al. 2006, 2008c).

Based on the above, it is clear that the periconception microenvironment of the oocyte affects its developmental competence and thus the subsequent pregnancy success. Whether it can also induce pertinent changes in the offspring's metabolism and body functions, and hence has an effect on its health later in life, is still a matter of debate. Farmers can influence this periconception environment by adequate measures (such as nutrition, health programs, and breeding technologies), as reviewed by Santos et al. (2010). It has been demonstrated that the altered microenvironment gives rise to altered patterns of gene expression in the offspring (Lillycrop and Burdge 2012). Epigenetic phenomena such as DNA methylation or histone modification are crucial in the regulation of gene expression, and intense epigenetic modifications have been shown to take place at a very high rate both in germ cells and in the preimplantation embryo. However, the question still remains whether the above-mentioned changes, such as high NEFA levels, which affect the microenvironment of the oocyte and later of the young embryo, can elicit epigenetic changes like DNA methylation or histone modifications.

3.3 *Nutrition*

In dairy cattle, the potential influence of nutrition on the periconception environment of the gamete and young embryo can be evaluated at three different levels: undernutrition, overnutrition, and diet composition.

3.3.1 Undernutrition

In modern dairy cows, undernutrition as such should be considered as a very rare or even nonexistent phenomenon, since animals that are to be inseminated (both nulliparous heifers and lactating multiparous cows) are generally fed according to their requirements. In beef cattle held under extensive conditions, undernutrition may, however, still occur, especially in specific seasons when animals are outdoors and the development of crops and grass is far below what is needed for animal feeding. Therefore, in lactating dairy cows, undernutrition is mainly regarded as incompetence to cope with the negative energy balance (NEBAL) during the immediate postpartum period. As mentioned earlier, the main challenge for the cows at that time is to optimize their dry matter intake in order to let the NEBAL not become too serious nor to last exceptionally long. As outlined by LeBlanc (2010) and (Mulligan et al. 2006), an inadequate management is the main underlying reason why cows fail to handle the NEBAL and finally experience severe metabolic stress. All too often, the latter leads to subclinical metabolic disease like subclinical ketosis or eventually even clinical ketosis and fatty liver.

Undernutrition in the postpartum dairy cow should therefore be seen as insufficient dry matter intake leading to inabilities for the cows to cope with NEBAL. In terms of reproduction, the latter will first of all be accompanied by a delayed resumption of normal ovarian cyclicity. Modern high-yielding dairy cows that do experience NEBAL have been shown to resume ovarian activity significantly later in comparison to cows in which the NEBAL was lower. Furthermore, significantly, more ovarian disturbances have been demonstrated in modern high-yielding dairy cows (Opsomer et al. 1998, 2000). In addition, the expression of heat symptoms has been shown to be significantly lower in those cows. The latter often necessitates farmers to inseminate cows based on secondary heat symptoms (such as restlessness and a decreased feed intake and milk yield), which are known to be associated with lower pregnancy results.

Most dairy cows develop the first postpartum dominant follicle approximately within 2 weeks after calving, but only about 40% of these follicles produce sufficient estradiol to stimulate ovulation, despite having normal ultrasound appearance and growth (Cheong et al. 2015). The mechanism leading to a correctly timed ovulation of a fertile oocyte is based on well-orchestrated cross talk within the hypothalamic-pituitary-ovarian axis (Cheong et al. 2015). In dairy cows selected for a high level of milk production, peripheral levels of glucose, insulin, and insulin-like growth factor-1 (IGF-1) are known to be substantially reduced (Marett et al. 2015; Bossaert et al. 2008). Lower peripheral insulin levels have been associated with non-ovulation of the dominant follicle, finally giving rise to cystic ovarian disease (Vanholder et al. 2005a). The underlying reason has shown to involve compromised theca cell function, finally leading to estradiol levels that are inadequate to provoke an ovulatory LH peak and hence ovulation. The reason for the compromised theca cell function has been attributed to the elevated levels of non-esterified fatty acids concomitant with overall fat mobilization (Vanholder et al. 2005b, 2006), although not all authors support this hypothesis. Overall, Cheong et al. (2015) recently concluded that cows that fail to ovulate the first postpartum dominant follicle are characterized by lower periparturient energy balance, increased insulin resistance, lower LH pulsatility, and lower intrafollicular concentrations of androstenedione and estradiol.

Undernutrition or, particularly for the dairy cow, insufficient dry matter intake and thus a more extensive NEBAL may cause adverse changes of metabolites in the ovarian follicular fluid. Although altered metabolite concentrations give rise to lower oocyte quality and hence lower fertilization rates as outlined above, this is less important for embryo growth because of the limited nutrient requirements of the early embryo and fetus for growth and development during the complete first half of gestation. It is furthermore well known that 75% of the growth of the ruminant fetus occurs during the last 2 months of gestation (Robinson et al. 1977). However, it is during the early phase of fetal development that maximal placental development and growth, differentiation, and vascularization occur, as well as fetal organogenesis, all of which being critical events for normal conceptus development. Nutrient restriction for the fetus is broadly defined as any series of events that reduce fetal and/or perinatal nutrient supply during critical windows of development. Basically, nutrient restriction can result from altered maternal nutrient supply, placental insufficiency, deranged metabolism and regulation, physiological extremes, and environmental conditions. From a practical standpoint, maternal nutrient supply and environmental conditions leading to stress responses are the most likely observed causes of nutrient restriction in ruminant livestock (Reynolds et al. 2006).

3.3.2 Overnutrition

Excessive energy intake particularly from high carbohydrate diets in cattle can reduce fertilization and embryo quality in some, but not all, circumstances. The latter has been attributed to increased circulating insulin levels during the final week of follicle growth, although the underlying mechanisms are still not clear (Wiltbank et al. 2014). Adamiak et al. (2005) demonstrated that high feeding levels were beneficial to nulliparous heifers in low body condition, but detrimental to oocytes from animals of moderately high body condition. Also here, elevated levels of insulin were considered as the underlying reason for this negative influence. Later, these findings were confirmed by Rooke et al. (2009) by feeding heifers diets high in starch, although they were able to avoid the adverse effects on oocyte quality when leucine intake was increased.

3.3.3 Composition of the Diet

Management strategies for transition cows (i.e., 3 weeks before calving until 2 weeks after calving) are mainly focused on helping the cows to cope with the metabolic load by optimizing health, minimizing stress (e.g., by minimizing the changes in group or ration), and stimulating dry matter intake and immune function. These are great opportunities for the veterinary practitioner to regularly monitor and adapt the herd management (Mulligan et al. 2006; LeBlanc 2010). A relatively new approach is to implement rather short-term changes in the quantity or composition of the diet at key stages in the reproductive process. Therefore, the term focus feeding, which refers to implementing short periods of nutritional supplements that are precisely timed and specifically designed to ameliorate the reproductive process including embryonic and fetal growth and development, has been introduced (Martin and Kadokawa 2006). In this context, Wiltbank et al. (2014) discussed possibilities to supplement rumen-protected fats in the ration of dairy cows. Of special interest herein is the supplementation of methionine since this is a rate-limiting amino acid for milk production and is known to be a methyl donor, which may potentially affect the status of DNA methylation and hence the expression level of certain genes.

Furthermore, application of diets specifically designed to improve fertility by counteracting mechanisms related to the NEBAL or by supporting a specific pathway necessary for successful fertility has always been a very attractive way to circumvent the impairment of reproduction during early lactation. Although the reproductive system is known to be influenced by multiple hormones that are also involved in the adaptation toward high milk production (growth hormone, IGF-1 and leptin), only insulin is known to be relatively sensible to the composition of the ration. Ovarian follicles have been shown to bear insulin receptors (Bossaert et al. 2010), and cows with lower peripheral insulin levels in the immediate postpartum period have been demonstrated to suffer from delayed postpartum ovarian resumption and are at higher risk to suffer from cystic ovarian disease (Vanholder et al. 2005a). Therefore, glucogenic diets have been advocated in the immediate postpartum period aiming to enhance the peripheral insulin concentrations and advance normal ovarian resumption (Gong et al. 2002). However, insulin has been shown to have detrimental effects on oocyte and embryo competence (Fouladi-Nashta et al. 2005) and to stimulate enzymatic catabolism of progesterone in the liver (Lemley et al. 2008). The latter suggests that glucogenic diets are only of advantage when offered in the immediate postpartum period and are to be avoided when cows are inseminated. In addition, low saturated fat diets around conception, reducing insulin levels, lead to increased conception rates (Garnsworthy et al. 2009).

Soybean meal contains isoflavones in concentrations that are able to induce increases in the blood concentration of estrogenically active isoflavone metabolites (equol, O-desmethylangolensin, dihydrodaidzein) in high-yielding dairy cows postpartum, even when supplemented in relatively low amounts (1.72 kg per day on average) (Cools et al. 2014). When compared with rapeseed meal, soy supplementation was furthermore associated with a decreased angio- and steroidogenesis at the level of the corpus luteum (CL) based on biopsy sampling at day 9 of the estrous cycle. However, no effect on the peripheral progesterone concentration during the first three estrous cycles after calving could be demonstrated. Therefore, although the results of the study suggest negative effects of soy feeding on CL function in recently calved dairy cows, the contribution of this effect on the peripheral progesterone concentration and consequently on overall fertility of supplemented cows warrants further research (Cools et al. 2014).

Adding fats to the diet has been another strategy that has been extensively tested to reduce the impaired reproductive capacity of dairy cows. In a study aiming to minimize the negative energy balance by decreasing the milk fat synthesis, and hence limiting energy output via milk by supplementing the ration with exogenous fats, we were not successful since cows simply produced more milk when reducing the NEBAL (Hostens et al. 2011). Omega-6 fatty acids are believed to have pro-inflammatory, and thus prostaglandin F2α-stimulating, properties rendering them of extra value early postpartum, while omega-3 fatty acids can weaken this inflammatory potency, leading to a higher chance of survival of the embryo when supplemented during the periconception period. However, the consequences of these fat feeding strategies on oocyte and embryo quality remain an intriguing issue for debate. Fat feeding may alter the microenvironment of the growing and maturing oocyte of the early and older embryo and thus may affect reproductive outcome. Dietary-induced hyperlipidemic conditions can thus also be harmful for embryo development and metabolism (for review, see Leroy et al. 2008a, b, c, 2014). Furthermore, peripheral blood in lactating dairy cows will contain a mixture of fatty acids of dietary origin and from body tissue breakdown, the latter being largely abundant in the immediate postpartum period and containing a high proportion of saturated fatty acids. Especially, the latter have been shown to have a significantly detrimental effect on both the oocyte and embryo quality (Leroy et al. 2005).

Supplementation of extra vitamins and minerals to the diet has often been suggested as the golden bullet solution to reduce the fertility decline. Usually, farmers are highly sensitive to this kind of advice since it does not involve extra labor, which is their paramount constraint nowadays. In herds in which cows are given a high amount of concentrates to sustain peak yield during the immediate postpartum period, the risk of suffering from such deficiencies is lower due to the fact that concentrates are usually highly supplemented with vitamins and minerals. In terms of their effect on immune response and embryo quality, special attention should be given to vitamin E and selenium. The latter was supported by the recent finding that during the dry period, treatment of tocopherol-deficient cattle with injectable vitamin E of 1000 IU each week for the last 3 weeks of gestation not only reduced the incidence of retained placenta and stillbirth but also significantly decreased pregnancy loss (20.5% vs. 12.5%; $P < 0.01$) (Pontes et al. 2015).

3.4 Heat Stress

Embryos are known to be very sensitive to the transient increases in body temperature arising as a result of elevated environmental temperature (heat stress), and dairy cows are very susceptible to heat stress since increasing milk yields interfere with body temperature regulation during warm weather, further exacerbating the deleterious effects on fertility. Heat stress is known to affect many components of the reproductive system including gonadotrophin profiles, follicular growth, granulosa cell function, steroidogenesis, and oocyte and embryo quality (for review see Roth 2008). Interestingly, observations of impaired fertility of dairy cattle in the autumn subsequent to a hot summer have been reported. The latter suggests a clear carry-over effect of heat stress. It seems that heat stress not only affects antral follicles emerging in the follicular wave but probably also affects the ovarian pool of small antral follicles resulting in a carry-over effect on follicular function and oocyte developmental competence.

3.5 Health Problems and Inflammatory Reactions

Forty to seventy percent of dairy cows across different levels of milk production, breeds, and management systems develop metabolic or infectious diseases in the immediate postpartum period (Dobson et al. 2007; Ribeiro et al. 2013). The calving-to-pregnancy interval is extended for at least 7, 8, 26, and 31 days in cows treated for mastitis, retained fetal membranes, hypocalcemia, and endometritis, respectively, compared with healthy herd-mates. Lameness is associated with even worse reproduction performance, as up to 40 days can be lost to get lame cows in-calf again even though the lameness has been treated (Dobson et al. 2007). In part, these poor fertility data may be related to delayed resumption of ovarian cyclicity after calving and on a lowered expression of heat symptoms. On the other hand, some events seem to have more long-lasting effects. Signs of dystocia or immediate postpartum hypocalcemia, endometritis, or mastitis can be "cured" within days by clinical treatment, but the cows are subfertile many weeks later during the breeding period. Obviously, inflammatory diseases taking place in the first weeks of lactation are associated with a reduced fertilization of cows inseminated between 50 and 60 days postpartum (Santos et al. 2010).

In a recent study, Ribeiro et al. (2016) showed that the carry-over effects of disease on reproduction of dairy cows cannot be explained simply by the nutritional status and its consequences to body condition score and estrous cyclicity at the onset of breeding postpartum. The inflammatory mediators produced by the injured or infected tissues can reach the reproductive tract including ovaries and uterus, but also the brain, which ultimately affects the physiological processes that control normal reproductive cyclicity. For example, cows that suffered from uterine disease postpartum had delayed growth of the first dominant follicle postpartum and reduced concentrations of estradiol (Sheldon et al. 2002). The presence of lipopolysaccharide (LPS; i.e., an endotoxin) in the follicular fluid of cows with uterine diseases has been postulated as a potential reason for compromised steroidogenesis, follicle growth, and impaired oocyte developmental competence (Bromfield et al. 2015).

3.6 Timing of Insults Affecting Reproductive Success and Embryonic Development

More and more evidence is currently indicating that the periconception environment is not only decisive for the fertility of the cow on the short term, but may be as important for general health of the offspring both in the immediate postnatal period and in later life. Specific approaches to improve management of dairy cattle during the periconception period and during pregnancy may therefore not only enhance the reproductive success of the dam but also the growth potential, health, and performance of her offspring later in life. If such innovative approaches could be available for use on the farm, producers might be able to increase animal health concomitantly with an improvement of the quality of the product while reducing costs of production.

The time at which insults are exposed upon dairy cattle will definitely influence reproductive outcome. During the earliest stages of pregnancy, the nutrient requirements of the embryo and young fetus are considered to be very low, causing undernutrition, for example, to be of lesser importance at that stage. Later, in the first trimester, however, development of specific organ systems (e.g., mammary gland, ovaries, liver, and pancreas) takes place. Hence, deleterious insults taking place at that time during pregnancy might be associated with impaired production and reproductive capacities and a lowered ability to ensure homeostasis in later life. During the second trimester of pregnancy, the fetus continues to develop and grow but, at the end of this stage, will still only reach about 25% of its size at birth. Therefore, and because the dam in most cases won back positive energy balance at that time, the risk of major metabolic challenges is lower during that stage. On the other hand, the development of major organ systems is still going on, so that major insults might still have pronounced effects on later health and productivity. The largest increase in fetal tissue size (75% of fetal growth) generally takes place during the final trimester of pregnancy, insults taking place at that time being mostly reflected in a significantly lowered birth weight.

4 Factors Affecting Periconception Environment in Assisted Reproductive Techniques, Such as Artificial Insemination and Embryo Transfer

4.1 Artificial Insemination (AI)

Artificial insemination (AI) consists of the introduction of sperm into the female genital tract for the purpose of achieving a pregnancy. At present, AI is a mature technique in dairy cattle breeding and established worldwide. Moreover, it yields pregnancy rates similar to natural mating. In a large study comparing AI with natural mating, no difference was found in fertility between both groups (Buckley et al. 2003). Among the adjustment variables in the model, those significantly associated with the likelihood of conception rate to first service included the herd, calving period, calving to first-service interval, and peak milk yield (as discussed above).

In general, frozen semen is used for AI in cattle; semen samples are thawed just before use and are deposited into the corpus uteri. Sperm numbers introduced vary around ten million live sperm cells (20 million frozen per straw), but it has been shown that numbers as low as two million give equally good pregnancy rates (49–53%) and that deep insemination into the uterine horn is not necessary (Verberckmoes et al. 2005). Deep insemination with normal dose of semen did not improve fertilization rates or embryo production in superovulated cattle (Carvalho et al. 2013). In one study in beef cattle, however, deep insemination gave better results (67% vs 49%) with low-dose (four million) semen than conventional insemination (Meirelles et al. 2012). Deep insemination of frozen-thawed semen into the uterine horns gives also better results in pigs (Vazquez et al. 2008), and in sheep, laparoscopic insemination with frozen-thawed semen deposition into the uterine horns has been the routine procedure for many decades, since vaginal or even deep cervical insemination is yielding much lower results (Evans 1988).

Detection of estrus is very important when performing AI, and results can easily be influenced by this aspect. In order to make the timing of artificial insemination (AI) relative to ovulation less critical, methods for prolonging shelf life of spermatozoa in vivo after AI have been developed. Encapsulation of sperm cells is a documented technology, and recently, a technology in which sperm cells are embedded in alginate gel has been introduced and commercialized. A blind field trial has been performed in Norway using standard processed semen with the Biladyl extender (control) in comparison with semen processed by sperm immobilization technology developed by SpermVital AS (Standerholen et al. 2015). Here it was demonstrated that fertility was not affected by encapsulation (NRR in both groups of 73%), although higher percentages of acrosome-damaged sperm were present in the encapsulated group. This shows that AI is a mature technique and can be used with different types of semen.

However, differences in bull fertility remain a problem. Nowadays, differences between bulls in field fertility have been related to the presence of "compensable" and "uncompensable" effects (Saacke 2008). Males requiring more sperm to be fertile after AI are considered to have compensable seminal deficiencies. This was elegantly shown by Den Daas et al. (1998), in a field trial where it became clear that the minimum sperm numbers required to achieve maximum pregnancy rates differ between bulls. These so-called compensable seminal deficiencies include a number of known sperm viability and morphology traits preventing fertilization to occur at all. Differences in fertility among males independent of sperm dosage are considered "uncompensable." These seminal deficiencies are associated with fertilizing sperms that are incompetent to maintain the fertilization process or subsequent embryogenesis (once initiated), leading to early embryonic mortality (Saacke 2008). It is obvious that compensable differences between males are more important in the AI industry than in extensive breeding systems, since AI with low doses of semen is economically more profitable, and hence bulls with compensable traits, which do not perform well at low doses, are less attractive and therefore are in less demand. Also interesting though are the uncompensable traits: can they be induced by environmental influences, with stressors having deleterious effects upon spermatogonia of the young or even prepubertal bull, or even earlier, when a male fetus was being exposed in utero to an adverse periconception environment? This may seem far-fetched and has not been investigated in cattle yet, but in mice there have been several studies that show an effect of in utero exposure on future male fertility of offspring, even for several generations. In some recent animal studies, mostly in mice and rats, transgenerational inheritance of acquired traits has been demonstrated. One such study examined the effect of endocrine disruptors on pregnant rats and their offspring (Anway et al. 2005). To this end, rats were injected with vinclozolin (an antiandrogenic endocrine disruptor) at 8–15 days post-coitum, which is the period of gonadal sex determination. Transient exposure of a gestating female rat to endocrine disruptors induced an adult phenotype in the F1 generation of increased incidence of male infertility. These effects were transferred through the male germ line to nearly all males of all subsequent generations examined (i.e., F1 to F4) (Anway et al. 2005). The effects on reproduction correlate with altered DNA methylation patterns in the germ line (Anway and Skinner 2008). However, when a similar study was performed by orally exposing pregnant rats to vinclozolin, to test a more natural route, no effect on male fertility in F1 or subsequent generations was found (Schneider et al. 2008).

Similar data are more difficult to achieve in large animal models, because of the large generation interval. Most large animal studies focus on the F1 generation to evaluate an effect of the environment on the offspring – it is important that such studies are also being performed on bulls and other large male mammals.

At present, there is no indication that calves produced after AI are less viable or less fertile than calves produced after natural mating, but this has apparently not been investigated yet (no hits on PubMed with calves * natural mating * insemination). It has been argued however that sex ratio may be skewed after artificial insemination, leading to more male calves (Berry and Cromie 2007). This was not confirmed in another study performed in Ethiopia, where a normal sex ratio of 50:50 was found after AI, but with odds ratio 1.38 for female calves after natural mating (Delesa et al. 2014). This was probably due to the fact that cattle that came into estrus during the harsh season were more likely to give birth to a female calf, according to the Trivers and Willard hypothesis (Trivers and Willard 1973) which states that as maternal condition declines, the adult female tends to produce a lower ratio of males to females.

However, the perception that AI gives more males than natural mating, which was obvious from both studies, needs further investigation, since such assumptions may prevent AI from being introduced at a large scale in Africa, where female cattle are much preferred over male calves.

4.2 Multiple Ovulation and Embryo Transfer (MOET)

Multiple ovulation and embryo transfer has been used in farm animals since the early 1980s (Thibier 2005). It consists of the use of multiple hormone injections to induce the release of multiple oocytes from the ovaries (so-called superovulation), followed by AI, flushing of the embryos from the uterus and transfer of the resulting embryos into synchronized recipients. Therefore, this technique increased offspring from genetically valuable females, while reducing also the time between generations. The major disadvantage of MOET is the unpredictable response of the donor female to exogenous superovulation treatment: 20% do not react sufficiently, whereas in *Bos taurus* an average of five to six embryos can be expected.

Multiple ovulation or superovulation has been reported to induce changes in gene expression and DNA methylation patterns in several species. In addition, the use of high dosages of gonadotropins induced spindle and chromosomal abnormalities in bovine oocytes (Liu et al. 2011). In mice, *H19*, *Snrpn*, *Peg3*, and *Peg1* showed loss of DNA methylation in mouse blastocysts obtained after superovulation, and this alteration was dose dependent, with aberrant methylation more frequent at high hormone dosage (Market-Velker et al. 2010b). Importantly, the loss of DNA methylation of *H19*, *Snrpn*, and *Peg1* could be observed in the sperm of the mouse offspring during two generations (Stouder et al. 2009). Superovulation in mice also resulted in biallelic expression of *Snrpn* and *H19* imprinted genes in the placenta (Fortier et al. 2008).

4.3 *In Vitro Embryo Production (IVP)*

In vitro embryo production entails the combination of three steps that need to be performed following a strict timing. It covers all steps from the maturation and fertilization of the oocyte to the embryo development. The oocytes can be derived both from a living (by transvaginal ovum pickup) or dead animal (slaughterhouse-derived ovaries). It is also combined with embryo transfer in many species.

4.3.1 In Vitro Embryo Production Affects Embryo Development and Quality

Cleavage and blastocyst rates are the main noninvasive parameters to assess bovine embryo quality. However, over time, it has become evident that high cleavage/blastocyst rates do not necessarily correlate with excellent quality of IVP embryos. Therefore, other criteria have been introduced, to compare with in vivo-derived cattle embryos. It is clear that IVP embryos show a darker cytoplasm due to their higher lipid content (Pollard and Leibo 1994), a more fragile zona pellucida (Duby et al. 1997), differences in metabolism (Khurana and Niemann 2000), a reduced intracellular communication (Boni et al. 1999), higher incidence of chromosome abnormalities (Viuff et al. 1996; Lonergan et al. 2004), errors of imprinting (Doherty et al. 2000), slower growth rate, higher thermal sensitivity, lower ICM/TE cell ratio (Van Soom et al. 1997a, b), and differences in gene expression compared to their in vivo counterparts (Driver et al. 2012). Additionally, higher apoptotic rates have been reported in IVP embryos compared to their in vivo counterparts (Gjørret et al. 2003), with an increased incidence of apoptosis as the culture time increases (Vandaele et al. 2006). Surprisingly, in cattle, some media used for in vitro culture are reported to have an influence on the sex ratio of the produced embryos, with a shift toward male embryos (Massip et al. 1996; Gutierrez-Adan et al. 2001).

4.3.2 Effects of In Vitro Embryo Production on Gene Expression

Gene expression is definitely different in IVP embryos. Before the development of wide screening techniques (such as microarray and RNA-seq), the effort was focused on genes known to play important roles during pre- and postimplantation development. The expression of *DNMT1*, *3A*, and *3B* has been demonstrated to be upregulated in bovine oocytes after in vitro maturation (IVM) compared to in vivo-matured oocytes (Heinzmann et al. 2011). In vitro embryo culture was shown to have a major effect on gene expression, which is logical since embryos spend 7–8 days in that environment. A microarray study showed that approximately 85% of differentially expressed genes was downregulated in IVP bovine blastocysts compared to their in vivo counterparts (Corcoran et al. 2006). Most of these genes are involved in transcriptional and translational events suggesting that a deficient machinery associated with transcription and translation is behind the inferior quality of IVP embryos (Corcoran et al. 2006). Furthermore, different culture media had a different impact on genes associated with transcription and translation (Corcoran et al. 2007). This is not a unique case; genes involved in blastocyst formation such as cell-to-cell adhesion (E-cadherin, connexins, TJ genes), cell communication (gap junctions), differentiation marks (Miller et al. 2003; Lonergan et al. 2003a, b; Boni et al. 1999), and genes related to apoptosis and oxidative stress (*BAX*, *SOX*, *HSP70*) had different expression patterns between different culture media (Sagirkaya et al. 2006; Rizos et al. 2002).

When comparing in vivo-derived IVP bovine embryos, genes related to metabolism, growth, and differentiation (*GLUT-5*, *CX43*, *IGF-II*, *LIF*) were upregulated in embryos derived in vivo, while genes related to stress (*SOX*, *MnSOD*, *BAX*, *HSP70.1*, *PRDX5*) were upregulated in IVP embryos. The significant increase in expression of those genes supports the hypothesis that current in vitro culture systems are associated with a considerable amount of oxidative stress (Rizos et al. 2003; Lazzari et al. 2002). Additionally, in embryos produced in vitro in the absence of fetal bovine serum (FBS), the expression of genes involved in the cholesterol biosynthesis pathway was upregulated compared to in vivo-derived embryos (Driver et al. 2012). In addition to culture media composition, culture conditions such as oxygen concentrations were shown to have an impact on gene expression (Harvey et al. 2004).

Other ARTs also induce alterations on the transcriptome. Some studies reported differences in the mRNA expression profile of several genes between embryos produced with sex-sorted and unsorted semen (Morton et al. 2007), while other studies failed to find differences (Bermejo-Alvarez et al. 2010). Nevertheless, offspring from sex-sorted spermatozoa did not display more abnormalities than the controls (Seidel and Garner 2002). Vitrification of mouse oocytes arrested at MII induced downregulation of *Dnmt1*, *1o*, *3A*, *3B*, and *3 L* in MII and of *Dnmt3B* in blastocysts (reviewed by Anckaert and Fair 2015). Furthermore, blastocyst vitrification altered the microRNA transcriptome of mouse embryos. Four miRNAs (mmu-miR-199a-5p, mmu-miR-329-3p, mmu-miR-136-5p, and mmu-miR-16-1-3p) were upregulated, and one (mmu-miR-212-3p) was downregulated in vitrified compared to fresh mouse blastocysts (Zhao et al. 2015). Additionally, superovulation induced alterations in the gene expression of bovine oocytes (Chu et al. 2012). Despite all this, a similar expression of developmentally important genes was observed between in vivo and IVP embryos carried to term (Ghanem et al. 2011).

4.3.3 Effects of ARTs on Epigenetic Marks

The study of the effects of ARTs on the epigenetic pattern of the embryos and resulting offspring has gained more attention in recent years. IVP increased the levels of DNA methylation compared to embryos derived in vivo, in rats and mice (Zaitseva et al. 2007). In bovine blastocysts, IVP altered the DNA methylation profile, with longer in vitro culture being translated in higher alteration compared to in vivo-derived embryos (Salilew-Wondim et al. 2015). Cloned embryos also showed aberrant DNA methylation patterns in several species, including cattle (Dean et al. 2003) and sheep (Beaujean et al. 2004). However, in rabbits no differences in DNA methylation status were observed between cloned and IVP embryos (Shi et al. 2004).

Alterations of imprinting have been observed in embryos, placenta, and offspring produced by ARTs. A loss of DNA methylation in *Igf2R* and *Peg1* and a gain of methylation in *H19* were found after IVM compared to in vivo maturation in mice. This gain of methylation in *H19* was also reported in humans after IVM in five out of 20 oocytes analyzed (reviewed by Ventura-Juncá et al. 2015).

In the mouse, loss of DNA methylation was reported in *H19*, *Snrpn*, and *Peg3* in IVP embryos using different culture media (Market-Velker et al. 2010a; Doherty et al. 2000). Additionally, serum supplementation induced alterations on the DNA methylation pattern of various imprinted genes (*H19*, *Igf2*, *Grb7*, *Grb10*, and *Peg1*), faster rates of development, and long-term behavioral consequences in mouse embryos (reviewed by Velker et al. 2012a). Vitrification also altered the methylation status of imprinted genes, causing a loss of methylation in *H19* in murine embryos (Wang et al. 2010).

A condition of overgrowth can be induced by ART in ruminants and is referred to as large offspring syndrome (LOS). It is characterized by large size at birth, breathing difficulties, reluctance to suckle, and sudden perinatal death (Young et al. 1998). LOS is caused by the exposure of pre-elongation ruminant embryos to unusual environmental conditions. It is not exactly clear what environmental changes are important, but a major cause is the use of serum in the culture media (Sinclair et al. 1999). Recent studies provide evidence for epigenetic changes in LOS as this syndrome shows dysregulation of several similar imprinted genes such as *IGF2R*, *KCNQ1OT1*, or *CDKN1C* (Chen et al. 2013, 2015). Furthermore, loss of maternal-specific *SNRPN* methylation was found in the placentae from in vitro-fertilized and in vitro-cultured bovine embryos (reviewed by Velker et al. 2012b). We have recently used RNA sequencing to examine the effect of in vitro embryo production, in either serum-containing or serum-free media, on the global gene expression pattern of individual bovine blastocysts. Compared to in vivo-derived embryos, embryos produced in serum-containing medium had five times more differentially expressed genes than embryos produced in serum-free conditions (1109 vs. 207). Importantly, in vitro production in the presence of serum appeared to have a different impact on the embryos according to their sex, with male embryos having three times more genes differentially expressed than their female counterparts (1283 vs. 456). On the contrary, male and female embryos produced in serum-free conditions showed the same number (191 vs. 192) of genes expressed differentially;

however, only 44 of those genes were common in both comparisons. Interestingly, the pathways affected by in vitro production differed depending on the presence or absence of serum in the medium. For example, embryos produced in serum-containing conditions had a lower expression of genes related to metabolism, while embryos produced in serum-free conditions showed aberrations in genes involved in lipid metabolism (Heras et al. 2016).

5 Conclusion

In conclusion, the periconception environment plays an important role in natural and assisted breeding of cattle. Changes in nutrition, temperature, metabolic status, or health status during natural breeding can affect reproductive success and offspring health. Likewise, changes in medium composition or the presence of stressors to which gametes or embryos are exposed can also influence reproductive success and offspring health in assisted reproduction. Recent research in herd health management and follow-up on offspring will help us to improve management of high-producing dairy cattle, and recent advances of "omics" technologies may enable the application of new molecular methods for the identification of signaling molecules which are imposing epigenetic marks upon gametes and embryos during early life, thereby affecting also reproductive success and offspring health.

Acknowledgments The authors are members of the COST Action FA1201 Epiconcept.

References

Adamiak S, Mackie K, Watt R, Webb R, Sinclair K (2005) Impact of nutrition on oocyte quality: cumulative effects of body composition and diet leading to hyperinsulinemia in cattle. Biol Reprod 73(5):918–926. doi:10.1095/biolreprod.105.041483

Anckaert E, Fair T (2015) DNA methylation reprogramming during oogenesis and interference by reproductive technologies: studies in mouse and bovine models. Reprod Fertil Dev 27(5):739–754. doi:10.1071/RD14333

Anway MD, Cupp AS, Uzumcu M, Skinner MK (2005) Epigenetic transgenerational actions of endocrine disruptors and male fertility. Science 308(5727):1466–1469. doi:10.1126/science.1108190

Anway MD, Skinner MK (2008) Transgenerational effects of the endocrine disruptor vinclozolin on the prostate transcriptome and adult onset disease. Prostate 68(5):517–529. doi:10.1002/pros.20724

Banos G, Brotherstone S, Coffey MP (2007) Prenatal maternal effects on body condition score, female fertility, and milk yield of dairy cows. J Dairy Sci 90(7):3490–3499. doi:10.3168/jds.2006-809

Beaujean N, Taylor J, Gardner J, Wilmut I, Meehan R, Young L (2004) Effect of limited DNA methylation reprogramming in the normal sheep embryo on somatic cell nuclear transfer. Biol Reprod 71(1):185–193. doi:10.1095/biolreprod.103.026559

Bermejo-Alvarez P, Rizos D, Rath D, Lonergan P, Gutierrez-Adan A (2010) Sex determines the expression level of one third of the actively expressed genes in bovine blastocysts. Proc Natl Acad Sci 107(8):3394–3399. doi:10.1073/pnas.0913843107

Berry D, Cromie A (2007) Artificial insemination increases the probability of a male calf in dairy and beef cattle. Theriogenology 67(2):346–352. doi:10.1016/j.theriogenology.2006.08.003

Boni R, Tosti E, Roviello S, Dale B (1999) Intercellular communication in *in vivo*-and *in vitro*-produced bovine embryos. Biol Reprod 61(4):1050–1055. doi:10.1095/biolreprod61.4.1050

Bossaert P, De Cock H, Leroy J, De Campeneere S, Bols P, Filliers M, Opsomer G (2010) Immunohistochemical visualization of insulin receptors in formalin-fixed bovine ovaries post mortem and in granulosa cells collected *in vivo*. Theriogenology 73(9):1210–1219. doi:10.1016/j.theriogenology.2010.01.012

Bossaert P, Leroy J, De Vliegher S, Opsomer G (2008) Interrelations between glucose-induced insulin response, metabolic indicators, and time of first ovulation in high-yielding dairy cows. J Dairy Sci 91(9):3363–3371. doi:10.3168/jds.2008-0994

Bowley F, Green R, Amer P, Meier S (2015) Novel approaches to genetic analysis of fertility traits in New Zealand dairy cattle. J Dairy Sci 98(3):2005–2012. doi:10.3168/jds.2014-8266

Brickell J, Bourne N, McGowan M, Wathes D (2009) Effect of growth and development during the rearing period on the subsequent fertility of nulliparous Holstein-Friesian heifers. Theriogenology 72(3):408–416. doi:10.1016/j.theriogenology.2009.03.015

Britt JH (2008) Oocyte development in cattle: physiological and genetic aspects. Rev Bras Zootec 37(SPE):110–115. doi:10.1590/S1516-35982008001300013

Bromfield J, Santos J, Block J, Williams R, Sheldon I (2015) PHYSIOLOGY AND ENDOCRINOLOGY SYMPOSIUM: Uterine infection: linking infection and innate immunity with infertility in the high-producing dairy cow. J Anim Sci 93(5):2021–2033. doi:10.2527/jas.2014-8496

Buckley F, Mee J, O'Sullivan K, Evans R, Berry D, Dillon P (2003) Insemination factors affecting the conception rate in seasonal calving Holstein-Friesian cows. Reprod Nutr Dev 43(6):543–555. doi:10.1051/rnd:2004002

Burthe S, Butler A, Searle KR, Hall SJ, Thackeray SJ, Wanless S (2011) Demographic consequences of increased winter births in a large a seasonally breeding mammal (Bos taurus) in response to climate change. J Anim Ecol 80(6):1134–1144. doi:10.1111/j.1365-2656.2011.01865.x

Carvalho P, Souza A, Sartori R, Hackbart K, Dresch A, Vieira L, Baruselli P, Guenther J, Fricke P, Shaver R (2013) Effects of deep-horn AI on fertilization and embryo production in superovulated cows and heifers. Theriogenology 80(9):1074–1081. doi:10.1016/j.theriogenology.2013.08.008

Chen Z, Hagen DE, Elsik CG, Ji T, Morris CJ, Moon LE, Rivera RM (2015) Characterization of global loss of imprinting in fetal overgrowth syndrome induced by assisted reproduction. Proc Natl Acad Sci 112(15):4618–4623. doi:10.1073/pnas.1422088112

Chen Z, Robbins KM, Wells KD, Rivera RM (2013) Large offspring syndrome: a bovine model for the human loss-of-imprinting overgrowth syndrome Beckwith-Wiedemann. Epigenetics 8(6):591–601. doi:10.4161/epi.24655

Cheong SH, Sá Filho OG, Absalón-Medina VA, Pelton SH, Butler WR, Gilbert RO (2015) Metabolic and endocrine differences between dairy cows that do or do not ovulate first postpartum dominant follicles. Biol Reprod 114:18. doi:10.1095/biolreprod.114.127076

Chu T, Dufort I, Sirard M-A (2012) Effect of ovarian stimulation on oocyte gene expression in cattle. Theriogenology 77(9):1928–1938. doi:10.1016/j.theriogenology.2012.01.015

Cools S, Van Den Broeck W, Vanhaecke L, Heyerick A, Bossaert P, Hostens M, Opsomer G (2014) Feeding soybean meal increases the blood level of isoflavones and reduces the steroidogenic capacity in bovine corpora lutea, without affecting peripheral progesterone concentrations. Anim Reprod Sci 144(3):79–89. doi:10.1016/j.anireprosci.2013.12.008

Corcoran D, Fair T, Park S, Rizos D, Patel O, Smith G, Coussens P, Ireland J, Boland M, Evans A (2006) Suppressed expression of genes involved in transcription and translation in *in vitro* compared with *in vivo* cultured bovine embryos. Reproduction 131(4):651–660. doi:10.1530/rep.1.01015

Corcoran D, Rizos D, Fair T, Evans AC, Lonergan P (2007) Temporal expression of transcripts related to embryo quality in bovine embryos cultured from the two-cell to blastocyst stage *in vitro* or *in vivo*. Mol Reprod Dev 74(8):972–977. doi:10.1002/mrd.20677

Dean W, Santos F, Reik W (2003) Epigenetic reprogramming in early mammalian development and following somatic nuclear transfer. Semin Cell Dev Biol 14(1):93–100. doi:10.1016/S1084-9521(02)00141-6

Delesa EK, Yohannes A, Alemayehu M, Samuel T, Yehualaeshet T (2014) Calves' sex ratio in naturally and artificially bred cattle in central Ethiopia. Theriogenology 82(3):433–439. doi:10.1016/j.theriogenology.2014.04.027

Den Daas J, De Jong G, Lansbergen L, Van Wagtendonk-De Leeuw A (1998) The relationship between the number of spermatozoa inseminated and the reproductive efficiency of individual dairy bulls. J Dairy Sci 81(6):1714–1723. doi:10.3168/jds.S0022-0302(98)75739-x

Dobson H, Smith R, Royal M, Knight C, Sheldon I (2007) The high-producing dairy cow and its reproductive performance. Reprod Domest Anim 42(s2):17–23. doi:10.1111/j.1439-0531.2007.00906.x

Doherty AS, Mann MR, Tremblay KD, Bartolomei MS, Schultz RM (2000) Differential effects of culture on imprinted H19 expression in the preimplantation mouse embryo. Biol Reprod 62(6):1526–1535. doi:10.1095/biolreprod62.6.1526

Driver AM, Peñagaricano F, Huang W, Ahmad KR, Hackbart KS, Wiltbank MC, Khatib H (2012) RNA-Seq analysis uncovers transcriptomic variations between morphologically similar *in vivo*- and *in vitro*-derived bovine blastocysts. BMC Genomics 13(1):1. doi:10.1186/1471-2164-13-118

Duby R, Hill J, O'Callaghan D, Overstrom E, Boland M (1997) Changes induced in the bovine zona pellucida by ovine and bovine oviducts. Theriogenology 1(47):332. doi:10.1016/S0093-691X(97)82459-4

Erickson B (1966) Development and senescence of the postnatal bovine ovary. J Anim Sci 25(3):800–805. doi:10.2527/jas1966.253800x

Evans G (1988) Current topics in artificial insemination of sheep. Aust J Biol Sci 41(1):103–116

Fortier AL, Lopes FL, Darricarrère N, Martel J, Trasler JM (2008) Superovulation alters the expression of imprinted genes in the midgestation mouse placenta. Hum Mol Genet 17(11):1653–1665. doi:10.1093/hmg/ddn055

Fouladi-Nashta A, Gutierrez C, Garnsworthy P, Webb R (2005) Effects of dietary carbohydrate source on oocyte/embryo quality and development in high-yielding, lactating dairy cattle. Biol Reprod (special issue):135–136

Funston R, Deutscher G (2004) Comparison of target breeding weight and breeding date for replacement beef heifers and effects on subsequent reproduction and calf performance. J Anim Sci 82(10):3094–3099. doi:10.2527/2004.82103094x

Funston R, Martin J, Larson D, Roberts A (2012) Physiology and endocrinology symposium: nutritional aspects of developing replacement heifers. J Anim Sci 90(4):1166. doi:10.2527/jas.2011-4569

Garnsworthy P, Fouladi-Nashta A, Mann G, Sinclair K, Webb R (2009) Effect of dietary-induced changes in plasma insulin concentrations during the early post partum period on pregnancy rate in dairy cows. Reproduction 137(4):759–768. doi:10.1530/REP-08-0488

Gauly M, Bollwein H, Breves G, Brügemann K, Dänicke S, Daş G, Demeler J, Hansen H, Isselstein J, König S (2013) Future consequences and challenges for dairy cow production systems arising from climate change in Central Europe–a review. Animal 7(05):843–859. doi:10.1017/S1751731112002352

Ghanem N, Wondim DS, Tesfaye D, Gad AY, Phatsara C, Tholen E, Looft C, Schellander K, Hoelker M (2011) Bovine blastocysts with developmental competence to term share similar expression of developmentally important genes although derived from different culture environments. Reproduction 142(4):551–564. doi:10.1530/REP-10-0476

Gjørret JO, Knijn HM, Dieleman SJ, Avery B, Larsson L-I, Maddox-Hyttel P (2003) Chronology of apoptosis in bovine embryos produced *in vivo* and in vitro. Biol Reprod 69(4):1193–1200. doi:10.1095/biolreprod.102.013243

Gong J, Lee W, Garnsworthy P, Webb R (2002) Effect of dietary-induced increases in circulating insulin concentrations during the early postpartum period on reproductive function in dairy cows. Reproduction 123(3):419–427. doi:10.1530/reprod/123.3.419

González-Recio O, Ugarte E, Bach A (2012) Trans-generational effect of maternal lactation during pregnancy: a Holstein cow model. PLoS One 7(12):e51816. doi:10.1371/journal.pone.0051816

Gutierrez-Adan A, Lonergan P, Rizos D, Ward F, Boland M, Pintado B, De La Fuente J (2001) Effect of the *in vitro* culture system on the kinetics of blastocyst development and sex ratio of bovine embryos. Theriogenology 55(5):1117–1126. doi:10.1016/s0093-691x(01)00471-x

Hales CN, Barker DJ (2001) The thrifty phenotype hypothesis. Br Med Bull 60(1):5–20. doi:10.1093/bmb/60.1.5

Harvey A, Kind K, Pantaleon M, Armstrong D, Thompson J (2004) Oxygen-regulated gene expression in bovine blastocysts. Biol Reprod 71(4):1108–1119. doi:10.1095/biolreprod.104.028639

Heinzmann J, Hansmann T, Herrmann D, Wrenzycki C, Zechner U, Haaf T, Niemann H (2011) Epigenetic profile of developmentally important genes in bovine oocytes. Mol Reprod Dev 78(3):188–201. doi:10.1002/mrd.21281

Heras S, De Coninck DI, Van Poucke M, Goossens K, Pascottini OB, Van Nieuwerburgh F, Deforce D, De Sutter P, Leroy JL, Gutierrez-Adan A (2016) Suboptimal culture conditions induce more deviations in gene expression in male than female bovine blastocysts. BMC Genomics 17(1):1. doi:10.1186/s12864-016-2393-z

Hostens M, Fievez V, Vlaeminck B, Buyse J, Leroy J, Piepers S, De Vliegher S, Opsomer G (2011) The effect of marine algae in the ration of high-yielding dairy cows during transition on metabolic parameters in serum and follicular fluid around parturition. J Dairy Sci 94(9):4603–4615. doi:10.3168/jds.2010-3899

Ibánez L, Lopez-Bermejo A, Callejo J, Torres A, Cabré S, Dunger D, de Zegher F (2008) Polycystic ovaries in nonobese adolescents and young women with ovarian androgen excess: relation to prenatal growth. J Clin Endocrinol Metabol 93(1):196–199. doi:10.1210/jc.2007-1800

Khurana NK, Niemann H (2000) Energy metabolism in preimplantation bovine embryos derived *in vitro* or *in vivo*. Biol Reprod 62(4):847–856. doi:10.1095/biolreprod62.4.847

Knight C, Beever D, Sorensen A (2000) Metabolic loads to be expected from different genotypes under different systems. Br Soc Anim Sci Occasional Publication 24:27–36

Lazzari G, Wrenzycki C, Herrmann D, Duchi R, Kruip T, Niemann H, Galli C (2002) Cellular and molecular deviations in bovine *in vitro*-produced embryos are related to the large offspring syndrome. Biol Reprod 67(3):767–775. doi:10.1095/biolreprod.102.004481

LeBlanc S (2010) Monitoring metabolic health of dairy cattle in the transition period. J Reprod Dev 56(S):S29–S35. doi:10.1262/jrd.1056S29

Lemley C, Butler S, Butler W, Wilson M (2008) Short communication: insulin alters hepatic progesterone catabolic enzymes cytochrome P450 2C and 3A in dairy cows. J Dairy Sci 91(2):641–645. doi:10.3168/jds.2007-0636

Leroy J, Opsomer G, Van Soom A, Goovaerts I, Bols P (2008a) Reduced fertility in high-yielding dairy cows: are the oocyte and embryo in danger? Part I the importance of negative energy balance and altered corpus luteum function to the reduction of oocyte and embryo quality in high-yielding dairy cows. Reprod Domest Anim 43(5):612–622. doi:10.1111/j.1439-0531.2007.00960.x

Leroy J, Van Soom A, Opsomer G, Goovaerts I, Bols P (2008b) Reduced fertility in high-yielding dairy cows: are the oocyte and embryo in danger? Part II mechanisms linking nutrition and reduced oocyte and embryo quality in high-yielding dairy cows. Reprod Domest Anim 43(5):623–632. doi:10.1111/j.1439-0531.2007.00961.x

Leroy J, Vanholder T, Van Knegsel A, Garcia-Ispierto I, Bols P (2008c) Nutrient prioritization in dairy cows early postpartum: mismatch between metabolism and fertility? Reprod Domest Anim 43(s2):96–103. doi:10.1111/j.1439-0531.2008.01148.x

Leroy J, Sturmey R, Van Hoeck V, De Bie J, McKeegan P, Bols P (2014) Dietary fat supplementation and the consequences for oocyte and embryo quality: hype or significant benefit for dairy cow reproduction? Reprod Domest Anim 49(3):353–361. doi:10.1111/rda.12308

Leroy J, Vanholder T, Delanghe J, Opsomer G, Van Soom A, Bols P, Dewulf J, de Kruif A (2004) Metabolic changes in follicular fluid of the dominant follicle in high-yielding dairy cows early post partum. Theriogenology 62(6):1131–1143. doi:10.1016/j.theriogenology.2003.12.017

Leroy J, Vanholder T, Mateusen B, Christophe A, Opsomer G, de Kruif A, Genicot G, Van Soom A (2005) Non-esterified fatty acids in follicular fluid of dairy cows and their effect on developmental capacity of bovine oocytes *in vitro*. Reproduction 130(4):485–495. doi:10.1530/rep.1.00735

Leroy J, Vanholder T, Opsomer G, Van Soom A, Ad K (2006) The *In Vitro* development of bovine oocytes after maturation in glucose and β-hydroxybutyrate concentrations associated with negative energy balance in dairy cows. Reprod Domest Anim 41(2):119–123. doi:10.1111/j.1439-0531.2006.00650.x

Leroy JL, Valckx SD, Jordaens L, De Bie J, Desmet KL, Van Hoeck V, Britt JH, Marei WF, Bols PE (2015) Nutrition and maternal metabolic health in relation to oocyte and embryo quality: critical views on what we learned from the dairy cow model. Reprod Fertil Dev 27(4):693–703. doi:10.1071/RD14363

Lillycrop KA, Burdge GC (2012) Epigenetic mechanisms linking early nutrition to long term health. Best Pract Res Clin Endocrinol Metab 26(5):667–676. doi:10.1016/j.beem.2012.03.009

Liu S, Feng HL, Marchesi D, Chen Z-J, Hershlag A (2011) Effect of gonadotropins on dynamic events and global deoxyribonucleic acid methylation during in vitro maturation of oocytes: an animal model. Fertil Steril 95(4):1503–6.e1-3. doi:10.1016/j.fertnstert.2010.09.049

Lonergan P, Forde N (2015) The role of progesterone in maternal recognition of pregnancy in domestic ruminants. In: Regulation of implantation and establishment of pregnancy in mammals. Springer, Switzerland, pp 87–104. https://link.springer.com/chapter/10.1007/978-3-319-15856-3_6

Lonergan P, Gutiérrez-Adán A, Rizos D, Pintado B, De La Fuente J, Boland MP (2003a) Relative messenger RNA abundance in bovine oocytes collected *in vitro* or *in vivo* before and 20 hr after the preovulatory luteinizing hormone surge. Mol Reprod Dev 66(3):297–305. doi:10.1002/mrd.10357

Lonergan P, Rizos D, Gutierrez-Adan A, Moreira P, Pintado B, De La Fuente J, Boland M (2003b) Temporal divergence in the pattern of messenger RNA expression in bovine embryos cultured from the zygote to blastocyst stage *in vitro* or *in vivo*. Biol Reprod 69(4):1424–1431. doi:10.1095/biolreprod.103.018168

Lonergan P, Pedersen HG, Rizos D, Greve T, Thomsen PD, Fair T, Evans A, Boland MP (2004) Effect of the post-fertilization culture environment on the incidence of chromosome aberrations in bovine blastocysts. Biol Reprod 71(4):1096–1100. doi:10.1095/biolreprod.104.030635

Marett L, Auldist M, Moate P, Wales W, Macmillan K, Dunshea F, Leury B (2015) Response of plasma glucose, insulin, and nonesterified fatty acids to intravenous glucose tolerance tests in dairy cows during a 670-day lactation. J Dairy Sci 98(1):179–189. doi:10.3168/jds.2014-8205

Market-Velker B, Fernandes A, Mann M (2010a) Side-by-side comparison of five commercial media systems in a mouse model: suboptimal *in vitro* culture interferes with imprint maintenance. Biol Reprod 83(6):938–950. doi:10.1095/biolreprod.110.085480

Market-Velker BA, Zhang L, Magri LS, Bonvissuto AC, Mann MR (2010b) Dual effects of superovulation: loss of maternal and paternal imprinted methylation in a dose-dependent manner. Hum Mol Genet 19(1):36–51. doi:10.1093/hmg/ddp465

Martin GB, Kadokawa H (2006) "Clean, green and ethical" animal production. Case study: reproductive efficiency in small ruminants. J Reprod Dev 52(1):145–152. doi:10.1262/jrd.17086-2

Massip A, Mermillod P, Van Langendonckt A, Reichenbach H, Lonergan P, Berg U, Carolan C, De Roover R, Brem G (1996) Calving outcome following transfer of embryos produced *in vitro* in different conditions. Anim Reprod Sci 44(1):1–10. doi:10.1016/0378-4320(95)01467-5

Meirelles C, Kozicki LE, Weiss RR, Segui MS, Souza A, dos Santos IW, dos Santos Breda JC (2012) Comparison between deep intracornual artificial insemination (dIAI) and conventional artificial insemination (AI) using low concentration of spermatozoa in beef cattle. Braz Arch Biol Technol 55(3):371–374. doi:10.1590/S1516-89132012000300006

Miller DJ, Eckert JJ, Lazzari G, Duranthon-Richoux V, Sreenan J, Morris D, Galli C, Renard J-P, Fleming TP (2003) Tight junction messenger RNA expression levels in bovine embryos are dependent upon the ability to compact and *in vitro* culture methods. Biol Reprod 68(4):1394–1402. doi:10.1095/biolreprod.102.009951

Morton K, Herrmann D, Sieg B, Struckmann C, Maxwell W, Rath D, Evans G, Lucas-Hahn A, Niemann H, Wrenzycki C (2007) Altered mRNA expression patterns in bovine blastocysts after fertilisation *in vitro* using flow-cytometrically sex-sorted sperm. Mol Reprod Dev 74(8):931–940. doi:10.1002/mrd.20573

Mulligan F, O'Grady L, Rice D, Doherty M (2006) A herd health approach to dairy cow nutrition and production diseases of the transition cow. Anim Reprod Sci 96(3):331–353. doi:10.1016/j.anireprosci.2006.08.011

Opsomer G, Coryn M, Deluyker H, Ad K (1998) An analysis of ovarian dysfunction in high yielding dairy cows after calving based on progesterone profiles. Reprod Domest Anim 33(3–4):193–204. doi:10.1111/j.1439-0531.1998.tb01342.x

Opsomer G, Gröhn Y, Hertl J, Coryn M, Deluyker H, de Kruif A (2000) Risk factors for post partum ovarian dysfunction in high producing dairy cows in Belgium: a field study. Theriogenology 53(4):841–857. doi:10.1016/S0093-691X(00)00234-x

Pollard J, Leibo S (1994) Chilling sensitivity of mammalian embryos. Theriogenology 41(1):101–106. doi:10.1016/s0093-691x(05)80054-8

Pontes G, Monteiro P, Prata A, Guardieiro M, Pinto D, Fernandes G, Wiltbank M, Santos J, Sartori R (2015) Effect of injectable vitamin E on incidence of retained fetal membranes and reproductive performance of dairy cows. J Dairy Sci 98(4):2437–2449. doi:10.3168/jds.2014-8886

Rawlings N, Evans A, Chandolia R, Bagu E (2008) Sexual maturation in the bull. Reprod Domest Anim 43(s2):295–301. doi:10.1111/j.1439-0531.2008.01177.x

Reynolds LP, Caton JS, Redmer DA, Grazul-Bilska AT, Vonnahme KA, Borowicz PP, Luther JS, Wallace JM, Wu G, Spencer TE (2006) Evidence for altered placental blood flow and vascularity in compromised pregnancies. J Physiol 572(1):51–58. doi:10.1113/jphysiol.2005.104430

Ribeiro E, Gomes G, Greco L, Cerri R, Vieira-Neto A, Monteiro P, Lima F, Bisinotto R, Thatcher W, Santos J (2016) Carryover effect of postpartum inflammatory diseases on developmental biology and fertility in lactating dairy cows. J Dairy Sci 99(3):2201–2220. doi:10.3168/jds.2015-10337

Ribeiro E, Lima F, Greco L, Bisinotto R, Monteiro A, Favoreto M, Ayres H, Marsola R, Martinez N, Thatcher W (2013) Prevalence of periparturient diseases and effects on fertility of seasonally calving grazing dairy cows supplemented with concentrates. J Dairy Sci 96(9):5682–5697. doi:10.3168/jds.2012-6335

Rizos D, Gutierrez-Adan A, Perez-Garnelo S, De La Fuente J, Boland M, Lonergan P (2003) Bovine embryo culture in the presence or absence of serum: implications for blastocyst development, cryotolerance, and messenger RNA expression. Biol Reprod 68(1):236–243. doi:10.1095/biolreprod.102.007799

Rizos D, Lonergan P, Boland M, Arroyo-Garcia R, Pintado B, De La Fuente J, Gutierrez-Adan A (2002) Analysis of differential messenger RNA expression between bovine blastocysts produced in different culture systems: implications for blastocyst quality. Biol Reprod 66(3):589–595. doi:10.1095/biolreprod66.3.589

Robinson J, McDonald I, Fraser C, Crofts R (1977) Studies on reproduction in prolific ewes. J Agric Sci 88(03):539–552. doi:10.1017/S0021859600037229

Rooke J, Ainslie A, Watt R, Alink F, McEvoy T, Sinclair K, Garnsworthy P, Webb R (2009) Dietary carbohydrates and amino acids influence oocyte quality in dairy heifers. Reprod Fertil Dev 21(3):419–427. doi:10.1071/RD08193

Roth Z (2008) Heat stress, the follicle, and its enclosed oocyte: mechanisms and potential strategies to improve fertility in dairy cows. Reprod Domest Anim 43(s2):238–244. doi:10.1111/j.1439-0531.2008.01168.x

Saacke R (2008) Sperm morphology: its relevance to compensable and uncompensable traits in semen. Theriogenology 70(3):473–478. doi:10.1016/j.theriogenology.2008.04.012

Sagirkaya H, Misirlioglu M, Kaya A, First NL, Parrish JJ, Memili E (2006) Developmental and molecular correlates of bovine preimplantation embryos. Reproduction 131(5):895–904. doi:10.1530/rep.1.01021

Salilew-Wondim D, Fournier E, Hoelker M, Saeed-Zidane M, Tholen E, Looft C, Neuhoff C, Besenfelder U, Havlicek V, Rings F (2015) Genome-wide DNA methylation patterns of bovine blastocysts developed *In Vivo* from embryos completed different stages of development *In Vitro*. PLoS One 10(11):e0140467. doi:10.1371/journal.pone.0140467

Santos J, Bisinotto R, Ribeiro E, Lima F, Greco L, Staples C, Thatcher W (2010) Applying nutrition and physiology to improve reproduction in dairy cattle. Soc Reprod Fertil Suppl 67:387–403. doi:10.5661/rdr-vii-387

Schneider S, Kaufmann W, Buesen R, van Ravenzwaay B (2008) Vinclozolin—the lack of a transgenerational effect after oral maternal exposure during organogenesis. Reprod Toxicol 25(3):352–360. doi:10.1016/j.reprotox.2008.04.001

Seidel GE, Garner DL (2002) Current status of sexing mammalian spermatozoa. Reproduction 124(6):733–743. doi:10.1530/rep.0.1240733

Shahinfar S, Page D, Guenther J, Cabrera V, Fricke P, Weigel K (2014) Prediction of insemination outcomes in Holstein dairy cattle using alternative machine learning algorithms. J Dairy Sci 97(2):731–742. doi:10.3168/jds.2013-6693

Sheldon I, Noakes D, Rycroft A, Pfeiffer D, Dobson H (2002) Influence of uterine bacterial contamination after parturition on ovarian dominant follicle selection and follicle growth and function in cattle. Reproduction 123(6):837–845. doi:10.1530/rep.0.1230837

Shi W, Dirim F, Wolf E, Zakhartchenko V, Haaf T (2004) Methylation reprogramming and chromosomal aneuploidy in *in vivo* fertilized and cloned rabbit preimplantation embryos. Biol Reprod 71(1):340–347. doi:10.1095/biolreprod.103.024554

Sinclair KD, McEvoy T, Maxfield E, Maltin C, Young L, Wilmut I, Broadbent P, Robinson J (1999) Aberrant fetal growth and development after *in vitro* culture of sheep zygotes. J Reprod Fertil 116(1):177–186. doi:10.1530/jrf.0.1160177

Standerholen FB, Waterhouse KE, Larsgard AG, Garmo RT, Myromslien FD, Sunde J, Ropstad E, Klinkenberg G, Kommisrud E (2015) Use of immobilized cryopreserved bovine semen in a blind artificial insemination trial. Theriogenology 84(3):413–420. doi:10.1016/j.theriogenology.2015.03.028

Stouder C, Deutsch S, Paoloni-Giacobino A (2009) Superovulation in mice alters the methylation pattern of imprinted genes in the sperm of the offspring. Reprod Toxicol 28(4):536–541. doi:10.1016/j.reprotox.2009.06.009

Swali A, Wathes DC (2007) Influence of primiparity on size at birth, growth, the somatotrophic axis and fertility in dairy heifers. Anim Reprod Sci 102(1):122–136. doi:10.1016/j.anireprosci.2006.10.012

Tamminga S, Luteijn P, Meijer R (1997) Changes in composition and energy content of liveweight loss in dairy cows with time after parturition. Livest Prod Sci 52(1):31–38. doi:10.1016/S0301-6226(97)00115-2

Thibier M (2005) The zootechnical applications of biotechnology in animal reproduction: current methods and perspectives. Reprod Nutr Dev 45(3):235–242. doi:10.1051/rnd:2005016

Trivers RL, Willard DE (1973) Natural selection of parental ability to vary the sex ratio of offspring. Science 179(4068):90–92. doi:10.1126/science.179.4068.90

Van Soom A, Bols P, Boerjan M, Ysebaert M, de Kruif A (1997a) Morphology and/or hatching ability of *in vitro* produced bovine embryos is no reliable indicator of inner cell mass cell number. Theriogenology 47(1):302–302. doi:10.1016/s0093-691x(97)82429-6

Van Soom A, Ysebaert MT, de Kruif A (1997b) Relationship between timing of development, morula morphology, and cell allocation to inner cell mass and trophectoderm in *in vitro*-produced bovine embryos. Mol Reprod Dev 47(1):47–56. doi:10.1002/(sici)1098-2795(199705)47:1<47::aid-mrd7>3.0.co;2-q

Vandaele L, Mateusen B, Maes D, de Kruif A, Van Soom A (2006) Is apoptosis in bovine *in vitro* produced embryos related to early developmental kinetics and *in vivo* bull fertility? Theriogenology 65(9):1691–1703. doi:10.1016/j.theriogenology.2005.09.014

Vanholder T, Leroy J, Dewulf J, Duchateau L, Coryn M, Kruif AD, Opsomer G (2005a) Hormonal and metabolic profiles of high-yielding dairy cows prior to ovarian cyst formation or first ovulation post partum. Reprod Domest Anim 40(5):460–467. doi:10.1111/j.1439-0531.2005.00601.x

Vanholder T, Leroy J, Van Soom A, Opsomer G, Maes D, Coryn M, de Kruif A (2005b) Effect of non-esterified fatty acids on bovine granulosa cell steroidogenesis and proliferation *in vitro*. Anim Reprod Sci 87(1):33–44. doi:10.1016/j.anireprosci.2004.09.006

Vanholder T, Leroy JL, Van Soom A, Maes D, Coryn M, Fiers T, de Kruif A, Opsomer G (2006) Effect of non-esterified fatty acids on bovine theca cell steroidogenesis and proliferation *in vitro*. Anim Reprod Sci 92(1):51–63. doi:10.1016/j.anireprosci.2005.05.014

Vazquez JM, Roca J, Gil MA, Cuello C, Parrilla I, Vazquez JL, Martínez EA (2008) New developments in low-dose insemination technology. Theriogenology 70(8):1216–1224

Velker BAM, Denomme MM, Mann MR (2012a) Embryo culture and epigenetics. Methods Mol Biol 912:399–421. doi:10.1007/978-1-61779-971-6_23

Velker BAM, Denomme MM, Mann MR (2012b) Loss of genomic imprinting in mouse embryos with fast rates of preimplantation development in culture. Biol Reprod 86(5):143. doi:10.1095/biolreprod.111.096602

Ventura-Juncá P, Irarrázaval I, Rolle AJ, Gutiérrez JI, Moreno RD, Santos MJ (2015) *In vitro* fertilization (IVF) in mammals: epigenetic and developmental alterations. Scientific and bioethical implications for IVF in humans. Biol Res 48(1):1. doi:10.1186/s40659-015-0059-y

Verberckmoes S, Van Soom A, Dewulf J, Thys M, de Kruif A (2005) Low dose insemination in cattle with the Ghent device. Theriogenology 64(8):1716–1728. doi:10.1016/j.theriogenology.2005.04.017

Viuff D, Avery B, Greve T, King W, Hyttel P (1996) Transcriptional activity in *in vitro* produced bovine two-and four-cell embryos. Mol Reprod Dev 43(2):171–179. doi:10.1002/(sici)1098-2795(199602)43:2<171::aid-mrd6>3.0.co;2-o

Vonnahme KA, Evoniuk J, Johnson ML, Borowicz PP, Luther JS, Pant D, Redmer DA, Reynolds LP, Grazul-Bilska AT (2008) Placental vascularity and growth factor expression in singleton, twin, and triplet pregnancies in the sheep. Endocrine 33(1):53–61. doi:10.1007/s12020-008-9052-3

Wallace J, Luther J, Milne J, Aitken R, Redmer D, Reynolds L, Hay W (2006) Nutritional modulation of adolescent pregnancy outcome–a review. Placenta 27:61–68. doi:10.1016/j.placenta.2005.12.002

Wang Z, Xu L, He F (2010) Embryo vitrification affects the methylation of the H19/Igf2 differentially methylated domain and the expression of H19 and Igf2. Fertil Steril 93(8):2729–2733

Wathes D (2012) Mechanisms linking metabolic status and disease with reproductive outcome in the dairy cow. Reprod Domest Anim 47(s4):304–312. doi:10.1111/j.1439-0531.2012.02090.x

Wathes D, Pollott G, Johnson K, Richardson H, Cooke J (2014) Heifer fertility and carry over consequences for life time production in dairy and beef cattle. Animal 8(s1):91–104. doi:10.1017/S1751731114000755

Wiltbank M, Garcia-Guerra A, Carvalho P, Hackbart K, Bender R, Souza A, Toledo M, Baez G, Surjus R, Sartori R (2014) Effects of energy and protein nutrition in the dam on embryonic development. Anim Reprod 11(3):168–182

Wiltbank MC, Baez GM, Garcia-Guerra A, Toledo MZ, Monteiro PL, Melo LF, Ochoa JC, Santos JE, Sartori R (2016) Pivotal periods for pregnancy loss during the first trimester of gestation in lactating dairy cows. Theriogenology 86(1):239–253. doi:10.1016/j.theriogenology.2016.04.037

Wooding P, Burton G (2008) Comparative placentation: structures, functions and evolution. Springer Science & Business Media, Berlin, Heidelberg. doi:10.1007/978-3-540-78797-6

Wu G, Bazer FW, Wallace JM, Spencer TE (2006) Board-invited review: intrauterine growth retardation: implications for the animal sciences. J Anim Sci 84(9):2316–2337. doi:10.2527/jas.2006-156

Young LE, Sinclair KD, Wilmut I (1998) Large offspring syndrome in cattle and sheep. Rev Reprod 3(3):155–163. doi:10.1530/ror.0.0030155

Zaitseva I, Zaitsev S, Alenina N, Bader M, Krivokharchenko A (2007) Dynamics of DNA-demethylation in early mouse and rat embryos developed *in vivo* and *in vitro*. Mol Reprod Dev 74(10):1255–1261. doi:10.1002/mrd.20704

Zhao X, Hao H, Du W, Zhu H (2015) Effect of Vitrification on the MicroRNA Transcriptome in mouse blastocysts. PLoS One 10(4):e0123451. doi:10.1371/journal.pone.0123451

The Consequences of Maternal-Embryonic Cross Talk During the Periconception Period on Subsequent Embryonic Development

Dimitrios Rizos, Veronica Maillo, Maria-Jesús Sánchez-Calabuig, and Patrick Lonergan

Abstract The periconception period comprises the final maturation of sperm and the processes of fertilization and early embryonic development, which take place in the oviduct. The final goal of these important events is to lead to establishment of pregnancy leading to the birth of healthy offspring. Studies in rodents and domestic animals have demonstrated that environmental conditions experienced during early development affect critical aspects of future growth, metabolism, gene expression, and physiology. Similarly, in vitro culture of embryos can be associated with changes in fetal growth, gene expression and regulation, and postnatal behavior.

In the oviduct, the cross talk between the mother and gametes/embryo begins after ovulation, between the oocyte and the female reproductive tract, and continues with the sperm and the early embryo after successful fertilization. These signals are mainly the result of direct interaction of gametes and embryos with oviductal and endometrial cells, influencing the microenvironment at the specific location. Identifying and understanding the mechanisms involved in this cross talk during the critical period of early reproductive events leading to pregnancy establishment could potentially lead to improvements in current in vitro embryo production systems in domestic mammals and humans. In this review, we discuss current knowledge of the short- and long-term consequences of in vitro embryo production on embryo development.

Keywords Embryo development • Embryo quality • In vivo • In vitro • Embryo-maternal interaction

---

D. Rizos, B.Agr.Sc., PhD (✉) • V. Maillo • M.-J. Sánchez-Calabuig
Departamento de Reproducción Animal, Instituto Nacional de Investigación y Tecnología Agraria y Alimentaria (INIA), Ctra. de la Coruña Km. 5,9, 28040 Madrid, Spain
e-mail: drizos@inia.es

P. Lonergan
School of Agriculture and Food Science, University College Dublin, Belfield, Dublin 4, Ireland

A. Fazeli, W.V. Holt (eds.), *Periconception in Physiology and Medicine*, Advances in Experimental Medicine and Biology 1014,
DOI 10.1007/978-3-319-62414-3_4

1 Periconception Environment In Vivo

In vivo, mammalian oocytes and embryos develop in a complex and dynamic environment. First, in the ovarian follicle, the oocyte grows and matures, achieving full developmental competence (Fair et al. 1995). Prior to ovulation, the luteinizing hormone (LH) surge leads to the resumption of meiosis and final oocyte maturation. Oocyte maturation involves (i) nuclear maturation, i.e., progression from prophase I to metaphase II with extrusion of the first polar body, (ii) cytoplasmic maturation which involves organelle redistribution, and (iii) molecular maturation that involves the accumulation of specific mRNAs (Sirard 2001). It has been hypothesized that the quality of an oocyte is based on the presence of the appropriate set of mRNA and proteins stored during folliculogenesis (Wrenzycki et al. 2007). A defined oocyte-specific gene expression pattern arising during folliculogenesis is crucial for the acquisition of oocyte developmental competence; conversely, deficiencies in gene expression or dynamics that occur during follicle development may be linked to impaired oocyte competence (Eichenlaub-Ritter and Peschke 2002; Sirard et al. 2006).

After ovulation, in the oviduct, the oocyte undergoes fertilization and the first mitotic or cleavage divisions. Finally in the uterus, the blastocyst forms, hatches from the zona pellucida, and, depending on the species, either implants immediately [day 4.5 in rodents (Wang and Dey 2006) and days 6–10 in humans (Cha et al. 2012)], forms a large free-floating spherical structure which initiates implantation around day 40 (horses), or elongates and progressively attaches to the uterine wall (initiation between days 18–22 in cows, 15–18 in ewes, and 14–18 in pigs) (Senger 2003) (Fig. 1). These events must be properly orchestrated for successful pregnancy establishment and the delivery of a healthy offspring.

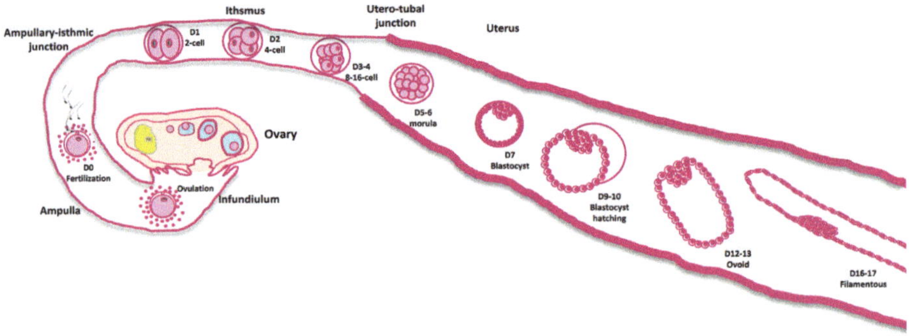

Fig. 1 Schematic representation of the process of early embryo development in vivo in cattle. After ovulation, the matured oocyte is fertilized (*D0*) at the ampullary-isthmic junction, while the first cleavage division takes place around 24–30 h later in the isthmus (*D1*) followed by subsequent mitotic divisions up to the 16-cell stage (*D4*). At this stage, the embryo passes into the uterus through the uterotubal junction and forms a morula (*D5–6*) and then a blastocyst (*D7*). After hatching from the zona pellucida (*D8*), the morphology of the embryo changes to ovoid (*D12–13*), then tubular and filamentous (*D16–17*) before implantation begins on D19

Following copulation, semen is deposited in the female reproductive tract which, depending on the species, will be in the cranial vagina (cow, sheep, primates, and cat) or into the uterine lumen (pig, horse, and dog) (Senger 2003). Typically, billions of sperm are deposited; however, during transit through the female reproductive tract, this number is dramatically reduced as sperm are selected by different barriers [cervix and uterotubal junction (UTJ)]. Once in the oviduct, the sperm are held in a storage reservoir in the isthmus which preserves sperm fertility, reduces the incidence of polyspermy by releasing sperm gradually (Suarez 2008), and constitutes the immediate source of viable sperm at the time of ovulation (Hunter and Wilmut 1984). It has been suggested that the sperm storage site recognizes and selects a fertile sperm population in mammalian species (Teijeiro and Marini 2012). After ovulation, the fimbria of the infundibulum that surrounds the ovary allows the passage of the ovulated oocyte into the oviduct. At this point, both muscle layers and ciliated cells mechanically guide the oocyte into the lumen of the ampulla to the site of fertilization (Hunter 1988). In addition, the oocyte loses the cumulus cells, and the zona pellucida (ZP) becomes exposed directly to the oviductal fluid (OF) which prepares it for fertilization and minimizes polyspermy (Coy et al. 2008, 2012).

Once fertilization occurs, the embryo spends the first 3–4 days in the oviduct, depending on the species. In cattle, the embryo remains in the oviduct until approximately day 4 of pregnancy, by which time it is at about the 16-cell stage. Then, the embryo enters the uterus and by day 7 forms a blastocyst consisting of an inner cell mass, which gives rise to the fetus, and the trophectoderm (TE), which forms the placenta. On days 9–10, the blastocyst hatches from the ZP and soon begins the process of elongation, which involves transitions from a spherical blastocyst on day 7 of gestation, through ovoid (days 12–13), tubular (days 14–15), and finally filamentous forms around days 16–17 (Fig. 1) (Degrelle et al. 2005). During elongation, the conceptus increases in size, more than 1000-fold, mainly through expansion of the trophectoderm (Betteridge et al. 1980) associated with an increase in protein content (Morris et al. 2000). After day 19, the elongated conceptus begins implantation with firm apposition and attachment of the trophectoderm to the endometrial luminal epithelium.

Proper communication between the conceptus-endometrium-corpus luteum (CL) is vital for pregnancy establishment. Progesterone (P4) synthesized by the CL acts indirectly via the endometrium to stimulate embryonic growth (Forde et al. 2009; Spencer et al. 2016). Conceptus elongation initiates interferon-τ (IFNT) production by TE cells (Roberts et al. 1999; Spencer and Bazer 2004; Robinson et al. 2008), which in cattle is the key signal for maternal recognition of pregnancy (Spencer and Bazer 2004). During maternal recognition of pregnancy, the mononuclear cells of the conceptus trophectoderm synthesize and secrete IFNT between days 10 and 21–25 with maximal production on days 14–16 (Bazer 1992; Roberts et al. 1999).

2 Periconception Environment In Vitro

In 1978, the first baby conceived by in vitro fertilization (IVF), Louise Brown, was born. Three years later, in 1981, the first calf obtained from IVF was born (Brackett et al. 1982). Although nearly 40 years have passed since then and many improvements have been made in in vitro embryo production (Paramio and Izquierdo 2016), even today in vitro systems are not as efficient as in vivo embryo production.

The goal of in vitro embryo production is to simulate as closely as possible the conditions that occur in vivo, to obtain high-quality embryos capable of continued development and implantation, and to result in viable births. Nowadays, in cattle, approximately 90% of oocytes cultured in vitro undergo nuclear and cytoplasmic maturation from which 80% are fertilized and cleave at least once (Lonergan et al. 2003a). Nevertheless, only between 30% and 40% reach the blastocyst stage (Rizos et al. 2008). In vitro, embryos are typically cultured until day 7 or 8 after fertilization, which corresponds to the blastocyst stage, when they are usually transferred into recipients. In heifers, the pregnancy rate following transfer of in vitro produced blastocysts is approximately 40–50% compared to about 70% when it comes to in vivo derived embryos (Hasler et al. 1995; Hoshi 2003). Thus, the challenge today is to improve current in vitro procedures providing high-quality embryos capable of continuing development and implantation after transfer to recipient and resulting in viable births.

2.1 *In Vitro Maturation and Effects on Oocyte Developmental Competence*

Cumulus-oocyte complexes (COCs) for research use are usually aspirated from ovaries recovered in the slaughterhouse, while those for commercial embryo transfer are recovered by transvaginal follicle puncture from live animals. In both cases, COCs are selected for in vitro maturation (IVM) based on morphological criteria as the compactness and thickness of cumulus and ooplasm homogeneity (Blondin and Sirard 1995). The cumulus cells play a critical role in the development of the oocyte by providing metabolites and nutrients, like pyruvate, oxaloacetic acid, and amino acids, thus stimulating them to resume meiosis and progress to metaphase II. Furthermore, cumulus cell expansion is an important marker for oocyte maturation, which is induced by gonadotrophin stimulation in vivo and in vitro leading to massive production of mucoid extracellular matrix protein (Chen et al. 1990). Recent evidence has shown that hyaluronic acid (HA), an important component of the extracellular matrix, plays an important role not only in cumulus expansion but also in oocyte maturation and further embryo development (Marei et al. 2012). Furthermore, transcriptome studies have identified a large number of genes in both oocytes and their associated cumulus cells that are involved in oocyte maturation (Regassa et al. 2011). Thus, the gene expression patterns in the cumulus cells have been used as an indicator of oocyte quality (Tesfaye et al. 2009; Bunel et al. 2015). Also, the detection of glucose-6-phosphate dehydrogenase activity by brilliant cresyl blue (BCB) staining can be used as a predictor of oocyte quality, where BCB-positive oocytes are more competent to form blastocyst than BCB-negative oocytes (34.1% vs. 3.9%, respectively, $P < 0.05$) (Alm et al. 2005).

Oocyte developmental competence, often defined as the ability of the oocyte to mature, be fertilized, and develop to the blastocyst stage, has been associated with (i) the size of the antral follicle from which it is recovered, (ii) the stage of the follicular wave, and (iii) the site of maturation – in vivo or in vitro [for review, see Lonergan and Fair 2016]. Oocytes matured in vivo are of better quality than those matured in vitro, and this is reflected in the number of embryos obtained subsequently. Indeed, it has been shown that irrespective of whether in vitro culture (IVC) occurred in vivo or in vitro, when oocytes were matured in vivo, the resultant blastocyst rate was almost 80%, while when oocytes were matured in vitro, it was limited to about 35% (Rizos et al. 2002). In relation to in vitro maturation, oocytes originating from follicles bigger than 6 mm resulted in significantly more blastocysts than those from 2 to 6 mm follicles and those recovered prior to the LH surge (Rizos et al. 2002).

The environment to which the oocyte is exposed during maturation can influence the abundance of transcripts in the matured oocyte (Watson 2000; Lonergan et al. 2003a) and in the resulting blastocyst (Knijn et al. 2002; Russell et al. 2006). Recent evidence has shown that suboptimal conditions during oocyte IVM have an effect at the epigenetic level and on genomic imprinting (Anckaert and Fair 2015) in bovine (Heinzmann et al. 2011) and ovine (Colosimo et al. 2009) oocytes. In vitro models that mimic the in vivo situation of high-yielding dairy cows after calving, i.e., in which nonesterified fatty acid (NEFA) concentrations are elevated, indicate that not only the oocyte developmental capacity is affected but also that the phenotype of the resulting embryos is altered (Van Hoeck et al. 2011, 2013). Maturing oocytes for 24 h under high saturated NEFA conditions significantly altered metabolic footprints of day 7 embryos at the level of both gene transcription and gene function (Van Hoeck et al. 2011). Nevertheless, additional studies are required to investigate whether the expression and DNA methylation of imprinted genes in blastocysts, fetuses, and placental tissue derived from IVM oocytes are unaltered [for review see (Lonergan and Fair 2016)].

Supplementation of putative growth-promoting substances to the maturation media (e.g., gonadotrophins, steroids, and growth factors) typically results in a modest improvement in the proportion of oocytes reaching the blastocyst stage (Thompson 2000). Normally, oocytes submitted to in vitro maturation are recovered from small- to medium-sized follicles (2–8 mm) which are capable of nuclear maturation but did not have sufficient time to undergo normal cytoplasmic maturation. A range of cellular and chemical methods have been successfully employed to artificially inhibit the meiotic resumption of oocytes following removal from the follicle, thereby allowing cytoplasmic maturation in vitro in the absence of nuclear maturation [for review, see Sirard 2001]. However, none of these in vitro approaches substantially improved oocyte developmental competence.

2.2 In Vitro Fertilization and Effects on Oocyte Developmental Competence

The fate of an embryo is determined at fertilization. Delays in fertilization or fertilization by a damaged spermatozoon could conceivably lead to oocyte aging or the formation of a defective embryo, respectively (Tarin et al. 2000). Any damage to the sperm after ejaculation can lead not only to a reduced fertilization rate but also to the formation of embryos with reduced ability to develop to the blastocyst stage. This phenomenon has been demonstrated for embryos formed from sperm exposed to gossypol (Brocas et al. 1997), oxidative stress (Silva et al. 2007), and sorting for gender by flow cytometry (Wheeler et al. 2006; Wilson et al. 2006; Bermejo-Alvarez et al. 2008; Bermejo-Alvarez et al. 2010). In addition, the nature of the sire itself can affect cleavage and the ability of the resulting embryos to develop to the blastocyst stage and to establish pregnancy after transfer [for review see (Hansen et al. 2010)]. Studies in trout have shown that the oocyte is able to partially repair sperm with damaged DNA during the first cleavage; however, when DNA repair is inhibited, damaged sperm is able to fertilize the oocyte but leads to embryo loss (Perez-Cerezales et al. 2010; Perez-Cerezales et al. 2011).

Cryopreserved semen is the main source of sperm for in vitro fertilization, although the proportion of fully functional sperm in a frozen-thawed sample is quite low (Holt 1997). Therefore, before in vitro fertilization, it is necessary to separate a motile fraction of sperm using one of a variety of methods such as centrifugation on a density gradient. During this process, seminal plasma is washed away and thereby much of the antioxidant protection is lost (Marques et al. 2010). In sperm, moderate levels of reactive oxygen species (ROS) are necessary for sperm maturation, capacitation, hyperactivation, acrosome reaction, and sperm-egg fusion (Kothari et al. 2010). Nevertheless, in vitro the sperm are often exposed to supraphysiological levels of ROS (Du Plessis et al. 2008) which can affect cell membranes, DNA, and mitochondria (Agarwal et al. 2006), leading to low fertilization rates and poor embryo quality (Silva et al. 2007; Jang et al. 2010). Antioxidants have been used to decrease the impact of oxidative stress, thereby improving sperm quality and their ability to fertilize an egg and consequently increasing the number of embryos obtained (Roca et al. 2004; Roca et al. 2005; Sapanidou et al. 2015) and their quality (Jang et al. 2010; Pang et al. 2016). However, other studies have failed to demonstrate any beneficial effects following the use of antioxidants during IVF and have even reported impaired fertilization and blastocyst rates (Ali et al. 2003; Goncalves et al. 2010; Marques et al. 2010). Therefore, more studies are needed to elucidate which antioxidants are best and when they should be added to improve the efficiency of IVF.

2.3 In Vitro Embryo Culture: Short- and Long-Term Effects

Embryo culture is the longest step during the process of in vitro embryo production and the step during which the greatest reduction in development occurs, achieving only 30–40% of blastocyst rate. The presumptive zygotes are recovered around 18 h

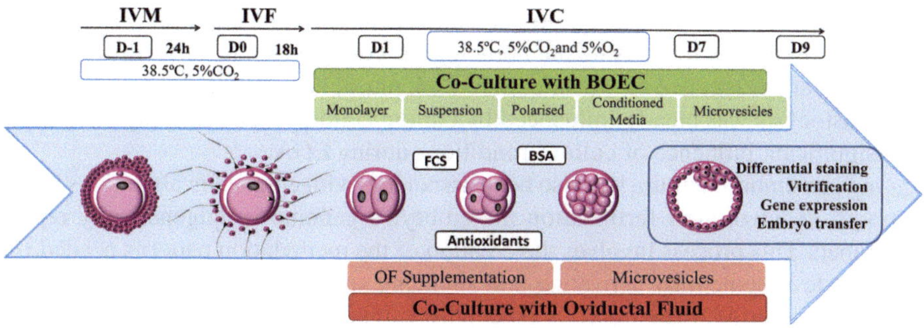

Fig. 2 Schematic representation of in vitro embryo production in cattle. In vitro maturation (*IVM*) of cumulus-oocyte complexes (COCs) is carried out for 24 h under an atmosphere of 38.5 °C, 5% CO_2, and maximum humidity. In vitro fertilization (*IVF*) is accomplished by co-incubation of matured COCs with selected motile sperm at a final concentration of 1×10^6 for 18 h, under the same conditions as IVM. In vitro culture (*IVC*) of the presumptive zygotes is maintained until D7–D9 at 38.5 °C, 5% CO_2, 5% O_2, and maximum humidity. The embryos can be cultured in different systems including coculture with BOECs in monolayers, suspension or polarized, and with products derived from BOECs such as microvesicles and conditioned media. Oviductal fluid recovered from the oviduct may also be used as well as its microvesicles

after fertilization and usually are cultured until days 7–9 (Fig. 2). It has been clearly shown that the intrinsic quality of the oocyte is the main factor affecting blastocyst yield, while the postfertilization culture environment affects the quality of the produced blastocysts (Rizos et al. 2002).

Culture conditions in vitro produce embryos of inferior quality compared to those derived from in vivo in terms of morphology, cryotolerance, transcript expression profiles, and pregnancy rate after transfer [for review, see Lonergan 2007]. To highlight the importance of developing in vitro systems that are as close as possible to the in vivo situation, different experiments have been carried out alternating in vitro with in vivo embryo culture. Thus, the culture of in vitro produced bovine zygotes in vivo in the sheep (Rizos et al. 2002), cow (Tesfaye et al. 2007), or mouse (Rizos et al. 2007) oviduct increases the quality of the resulting blastocysts so that it resembles that of totally in vivo produced embryos. Conversely, the in vitro culture of in vivo produced bovine zygotes results in blastocysts of low quality (Rizos et al. 2002). In addition, we have demonstrated that in vitro produced bovine zygotes cultured either in vitro or in vivo in the sheep oviduct at different stages of development up to the blastocyst stage exhibit a clear temporal sensitivity to their culture environment on their transcriptome and cryotolerance (Lonergan et al. 2003b). Thus, blastocysts cultured for the first 2 days in vivo and the last 4 days in vitro had the lowest survival rates after cryopreservation; those cultured only for the last 2 days in vitro had intermediate rates of survival, and those which spent the last 4 days of culture in vivo had high rates of survival, compared with those cultured entirely in vivo. Based on these studies, it was concluded that the period around the time of embryonic genome activation (EGA) is critical to the quality of the embryo. Similarly, Gad et al. (2012) investigated the consequences of culture conditions before and during the EGA on bovine embryonic developmental rates and global gene expression patterns using the homologous cow oviduct. Embryo development was similar irrespective of where culture took place; however, the transcriptome of the blastocysts was clearly influenced by culture conditions, confirming once more the significant influence of culture conditions during EGA.

In vitro embryo culture has also been associated with epigenetic alterations in the embryo. In vivo, after fertilization the embryo's genome is epigenetically reprogrammed. This process involves the creation of the methylation patterns needed for normal development by activating and silencing specific genes (Reik et al. 2001). During this period, the embryo is especially vulnerable to in vitro induced epigenetic defects (El Hajj and Haaf 2013). Indeed, animal studies have revealed a link between different assisted reproductive technologies (ARTs) and imprinting disorders, via altered DNA-methylation patterns and histone codes (Urrego et al. 2014). In addition, imprinting disorders are more prevalent in gametes and embryos after ARTs than in their counterparts derived from in vivo production (Urrego et al. 2014). In vitro embryo culture has been associated with abnormal reprogramming of imprinted genes such as *SNPRN*, *IGF2*, or *H19* in cattle (Nowak-Imialek et al. 2008; Curchoe et al. 2009; Suzuki et al. 2009) and mice (Khosla et al. 2001). These alterations to the epigenetic profile may have a direct effect on the subsequent embryo and fetal development.

From the above, it can be concluded that the actual in vitro culture conditions provide a suboptimal environment for early embryonic development and have short- and long-term consequences. The short-term consequences include alterations in morphology, cryotolerance, and gene expression patterns, while the long-term consequences are reflected in abnormal offspring development and behavior (Rizos et al. 2008).

Embryos can be cultured in defined or semi-defined media, cocultured with oviduct epithelial cells or their conditioned media and with extracellular vesicles (Maillo et al. 2016). Nowadays, one of the most commonly used media for the culture of bovine embryos is synthetic oviductal fluid (SOF) which is frequently supplemented with fetal calf serum (FCS) and/or bovine serum albumin (BSA) (Tervit et al. 1972; Holm et al. 1999). The presence of serum in the IVC media has a stimulatory effect on the speed of development, with more blastocysts on day 6 of culture than either in its absence or with BSA (Gutierrez-Adan et al. 2001; Rizos et al. 2003). However, serum can have a negative effect on embryo quality as manifested by reduced cryotolerance and altered gene expression (Lazzari et al. 2002; Rizos et al. 2003; Wrenzycki et al. 2005) and pregnancy rate after transfer. Lazzari et al. (2002) evidenced that IVC of bovine embryos in the presence of serum or BSA significantly increased the number of cells in day 7 blastocysts and the relative abundance of transcripts for several genes including *HSP70.1*, *Cu/Zn-SOD*, *GLUT-3*, *GLUT-4*, *bFGF*, and *IGFI-R* compared with embryos cultured in vivo (either in the sheep oviduct or produced entirely in vivo). Moreover, these deviations were linked to gestation length and birth weight of the derived calves. Both in vitro systems, FCS and BSA, were associated with a significantly elevated incidence of deviations in embryonic development and a higher proportion of calves with increased birth weight (Lazzari et al. 2002). In a recent study, it was shown that BSA removal over a 24 h period (from D6 to D7) in an individual embryo culture system decreased embryo development and cell counts in the inner cell mass, although the embryos tended to improve their survival after vitrification and also to result in a lower incidence of miscarriage (Murillo-Ríos et al. 2016).

Culture of embryos with FCS has been associated with alterations in the phenotype of newborn offspring in cattle (Farin et al. 2006), sheep (Walker et al. 1996), and mice (Fernandez-Gonzalez et al. 2004) named "large offspring syndrome (LOS)." LOS is characterized by aberrant placental development, extended gestation length, sudden perinatal death, breathing difficulties, more male calves, and large size at birth (Farin et al. 2006). Insulin-like growth factor 2 (IGF2) is an imprinted gene that regulates fetal and placental development in cattle and other species (Constancia et al. 2002). In sheep, the altered expression of *IGF2R* has been correlated with the incidence of LOS (Young et al. 2001) suggesting that IGF2 could be sensitive to epigenetic disorders during in vitro conditions. In addition, fetuses and placental tissue derived from IVP embryos presented an aberrant expression of imprinted and non-imprinted genes (Perecin et al. 2009; Farin et al. 2010). These phenotypic alterations may be the result of failure to properly establish or maintain DNA methylation and histone modifications during in vitro conditions (Farin et al. 2006). This also reflects the likelihood that epigenetic alterations during early embryo development are maintained in subsequent embryo and fetal development.

Oxidative stress is also important for the developing embryo during the embryo culture step. Substances such as ammonia, oxygen radicals, or growth factors can produce lipid peroxidation, membrane injury, and structural damage, leading to decreased cryotolerance and apoptosis (Somfai et al. 2007). Thus, antioxidant supplementation of the culture media has been shown to protect the embryo against oxidative stress. It also helps to maintain intra- and extracellular redox balance, which is necessary to reduce the toxicity of ROS, improving embryo development, increasing cryotolerance and cell differentiation, and inhibiting apoptosis during culture (Guerin and Menezo 2011; Takahashi 2012). Vitamins such as tocopherol, ascorbic acid, folic acid, or cyanocobalamin play a key role in reducing oxidative damage and improving blastocyst development rate in mouse embryos (Wang et al. 2002). Recently, it has been demonstrated that the supplementation of culture media with other antioxidants, such as cocretin which has a high antioxidant capacity, leads to an improvement of cryotolerance of bovine embryos (Zullo et al. 2016). Another antioxidant, retinol, seems to have a positive effect on early embryo development and quality (Livingston et al. 2004). Similar results were reported by Lawrence et al. (2004) suggesting that retinol protects early bovine embryos against damaging effect of heat stress during in vitro culture.

Despite the many efforts to improve in vitro culture media, conditions within the oviduct have not been fully recapitulated in vitro. As knowledge concerning embryo requirements during early embryonic development increases, in vitro culture systems have evolved to mimic more precisely what occurs in the oviduct during this period. Therefore, at this moment, it is known that the embryo requires an evolving array of energy substrates (Gardner et al. 1996; Quesenberry et al. 2015); consumption of pyruvate and glucose is low until the 16-cell stage and increases significantly with morula compaction and blastocyst formation (de Souza et al. 2015).

As previously mentioned, in an attempt to mimic in vivo conditions and improve the quality of the embryos produced, different systems of embryo culture have been developed. They offer unique advantages such as the gradual change of culture media to suit the specific requirements of the developing embryo, thus overcoming limitations of conventional culture systems. Coculture in vitro with bovine oviductal epithelial cells (BOECs) has been considered to help with the production of good quality embryos (Ulbrich et al. 2010). These cells can be grown as monolayers, cell suspensions, or as polarized cultures (Fig. 2). A study by Cordova et al. (2013) showed that the use of BOEC in in vitro embryo culture at the early stages of embryo development, up to day 4, improves embryo development and embryo quality in terms of expression of specific gene transcripts. This period of culture coincides with the in vivo period when the embryo is still in the oviduct. However, the drawback of monolayers is that they dedifferentiate, losing important morphological characteristics (Rottmayer et al. 2006) including reduction of cell height, loss of cilia, and loss of secretory granules and bulbous protrusions (Thibodeaux et al. 1992; Walter 1995). Recently, we demonstrated that an established BOEC monolayer can be used successfully for coculture with no differences in embryo development when compared either with coculture with fresh recovered cells or normal culture in SOF (Lopera-Vásquez et al. 2016). Furthermore, the quality of the produced blastocysts in terms of cryotolerance and number of TE and ICM cells was higher than those produced using media supplemented with FCS. This method gives an advantage over the classical coculture systems as it provides homogeneous results. An alternative to monolayers is the short-term (24 h) epithelial cell suspension culture, which maintains morphological characteristics as well as the gene markers present in the cell in vivo such as *OVGP1*, estrogen, and P4 receptors (Rottmayer et al. 2006). Preliminary results from our group showed that BOECs in suspension cultures are closer to in vivo controls than monolayers in terms of morphology and oviductal epithelial cell markers *OVGP1*, *GPX4*, and *FOXJ1* (Hamdi et al. 2015). Finally, polarized cell cultures maintain the polarized asymmetrical structure of the oviductal epithelial cells, and it seems that this system preserves detailed morphological features of the porcine oviduct as well as oviduct-specific markers (Miessen et al. 2011).

Oviductal fluid (OF) has been used as a supplement during in vitro embryo production. Its composition is very complex, containing simple and complex carbohydrates, ions, lipids, phospholipids, and proteins (Avilés et al. 2010). Porcine oocytes treated with OF before fertilization showed significantly increased cleavage rates and blastocyst yield, suggesting that OF protects the embryo against adverse effects on mitochondrial DNA transcription or replication and apoptosis (Lloyd et al. 2009). In contrast, Cebrian-Serrano et al. (2013) evidenced that the exposure of cattle oocytes to OF before fertilization had no effect on embryo development and morphology of the resulting blastocysts; however, blastocysts produced from oocytes treated with OF showed differences in specific transcripts (Cebrian-Serrano et al. 2013). Recently, we showed that low concentrations of OF (<5%) in bovine embryo culture media as a substitute for serum had a positive effect on development and quality in terms of cryotolerance, cell number, and expression of qualitatively related genes (Lopera-Vasquez et al. 2015).

Extracellular vesicles (EVs) is a general term encompassing several different vesicle types, released by somatic cells, that are present in body fluids and contain bioactive molecules (i.e., proteins, RNAs, mRNAs, and miRNAs) (Simons and Raposo 2009; Silveira et al. 2012) and lipids (Raposo and Stoorvogel 2013). The EV denomination is commonly size and origin associated, being exosomes (30–200 nm) from endosomal origin and microvesicles (MV) (100–1000 nm) from the plasma membrane. EVs are important in intercellular communication and play a key role in the regulation of physiological and pathological processes. It has been demonstrated that EV can horizontally transfer mRNAs to other cells, whereupon the mRNA can then be translated into functional proteins in the new location (Hergenreider et al. 2012). EVs have been identified in vivo in all body fluids including amniotic fluid, urine, and blood (Simpson et al. 2008). Currently, in reproduction, knowledge of the role of these secreted vesicles is limited to those from follicular fluid (Silveira et al. 2012), the endometrial environment (Ng et al. 2013), seminal plasma (Piehl et al. 2013), and uterine fluid (Burns et al. 2014). Recently, Burns et al. (2016) demonstrated that EVs emanate from both the conceptus trophectoderm and uterine epithelia and are involved in intercellular communication between those tissues during the establishment of pregnancy in sheep. Therefore, EV can be used as a supplement during in vitro embryo culture. Thus, Saadeldin et al. (2014) showed that the addition of exosomes isolated from the conditioned medium of parthenogenetic embryos increased the developmental competence of cloned embryos. In vitro derived embryos are also known to secrete EVs into culture media, where they may play a role in promoting development (Saadeldin et al. 2015). Recently, we provided evidence that EVs derived from BOEC conditioned media improve blastocyst quality and induce cryoprotection in in vitro cultures to the same extent as classical coculture with fresh BOEC monolayers (Lopera-Vásquez et al. 2016). Thus, the presence of EV in OF and their effect on early embryonic development may be of great importance and may provide information and new insights on early embryo-maternal communication and improve embryo quality in our current IVP systems.

3 Conclusion

During the last few decades, many advances have been achieved in our understanding of early reproductive events based on in vivo and in vitro studies, and assisted reproductive technologies are commonly used in humans and many animal species with success. Nevertheless, the quality of in vitro embryos is still inferior compared to their in vivo counterparts, and in vitro conditions have short- and long-term effects on the resulting embryo, fetus, and offspring. Therefore, a better understanding of how the embryo develops physiologically in the reproductive tract (oviduct and uterus) will provide the knowledge to develop new strategies to decipher the mechanisms involved in oocyte developmental competence that help and improve the current systems of in vitro embryo production. Together, such improvements will lead to the production of better quality embryos. Moreover, proper in vitro models that mimic the physiological situation as closely as possible will be developed.

Acknowledgements Funded by the Spanish Ministry of Economy and Competitiveness AGL2015- 70140-R. P.L. was supported by funding from the European Union Seventh Framework Programme FP7/2007–2013 under grant agreement n° 312097 ("FECUND").

References

Agarwal A, Said TM, Bedaiwy MA, Banerjee J, Alvarez JG (2006) Oxidative stress in an assisted reproductive techniques setting. Fertil Steril 86:503–512

Ali AA, Bilodeau JF, Sirard MA (2003) Antioxidant requirements for bovine oocytes varies during in vitro maturation, fertilization and development. Theriogenology 59:939–949

Alm H, Torner H, Lohrke B, Viergutz T, Ghoneim IM, Kanitz W (2005) Bovine blastocyst development rate in vitro is influenced by selection of oocytes by brilliant cresyl blue staining before IVM as indicator for glucose-6-phosphate dehydrogenase activity. Theriogenology 63:2194–2205

Anckaert E, Fair T (2015) DNA methylation reprogramming during oogenesis and interference by reproductive technologies: studies in mouse and bovine models. Reprod Fertil Dev 27:739–754

Avilés M, Gutiérrez-Adán A, Coy P (2010) Oviductal secretions: will they be key factors for the future ARTs? Mol Hum Reprod 16:896–906

Bazer FW (1992) Mediators of maternal recognition of pregnancy in mammals. Proc Soc Exp Biol Med 199:373–384

Bermejo-Alvarez P, Lonergan P, Rath D, Gutierrez-Adan A, Rizos D (2010) Developmental kinetics and gene expression in male and female bovine embryos produced in vitro with sex-sorted spermatozoa. Reprod Fertil Dev 22:426–436

Bermejo-Alvarez P, Rizos D, Rath D, Lonergan P, Gutierrez-Adan A (2008) Can bovine in vitro-matured oocytes selectively process X- or Y-sorted sperm differentially? Biol Reprod 79: 594–597

Betteridge KJ, Eaglesome MD, Randall GC, Mitchell D (1980) Collection, description and transfer of embryos from cattle 10–16 days after oestrus. J Reprod Fertil 59:205–216

Blondin P, Sirard MA (1995) Oocyte and follicular morphology as determining characteristics for developmental competence in bovine oocytes. Mol Reprod Dev 41:54–62

Brackett BG, Bousquet D, Boice ML, Donawick WJ, Evans JF, Dressel MA (1982) Normal development following in vitro fertilization in the cow. Biol Reprod 27:147–158

Brocas C, Rivera RM, Paula-Lopes FF, McDowell LR, Calhoun MC, Staples CR, Wilkinson NS, Boning AJ, Chenoweth PJ, Hansen PJ (1997) Deleterious actions of gossypol on bovine spermatozoa, oocytes, and embryos. Biol Reprod 57:901–907

Bunel A, Jorssen EP, Merckx E, Leroy JL, Bols PE, Sirard MA (2015) Individual bovine in vitro embryo production and cumulus cell transcriptomic analysis to distinguish cumulus-oocyte complexes with high or low developmental potential. Theriogenology 83:228–237

Burns G, Brooks K, Wildung M, Navakanitworakul R, Christenson LK, Spencer TE (2014) Extracellular vesicles in luminal fluid of the ovine uterus. PLoS One 9:e90913

Burns GW, Brooks KE, Spencer TE (2016) Extracellular vesicles originate from the conceptus and uterus during early pregnancy in sheep. Biol Reprod 94(3):56

Cebrian-Serrano A, Salvador I, Garcia-Rosello E, Pericuesta E, Perez-Cerezales S, Gutierrez-Adan A, Coy P, Silvestre MA (2013) Effect of the bovine oviductal fluid on in vitro fertilization, development and gene expression of in vitro-produced bovine blastocysts. Reprod Domest Anim 48:331–338

Cha J, Sun X, Dey SK (2012) Mechanisms of implantation: strategies for successful pregnancy. Nat Med 18:1754–1767

Chen L, Wert SE, Hendrix EM, Russell PT, Cannon M, Larsen WJ (1990) Hyaluronic acid synthesis and gap junction endocytosis are necessary for normal expansion of the cumulus mass. Mol Reprod Dev 26:236–247

Colosimo A, Di Rocco G, Curini V, Russo V, Capacchietti G, Berardinelli P, Mattioli M, Barboni B (2009) Characterization of the methylation status of five imprinted genes in sheep gametes. Anim Genet 40:900–908

Constancia M, Hemberger M, Hughes J, Dean W, Ferguson-Smith A, Fundele R, Stewart F, Kelsey G, Fowden A, Sibley C, Reik W (2002) Placental-specific IGF-II is a major modulator of placental and fetal growth. Nature 417:945–948

Cordova A, Perreau C, Schmaltz-Panneau B, Locatelli Y, Ponsart C, Mermillod P (2013) Use of an in vitro model in bovine to evidence a functional and molecular dialogue between preimplantation embryo and oviduct epithelial cells. Gynecol Obstet Fertil 41:537–539

Coy P, Cánovas S, Mondéjar I, Saavedra MD, Romar R, Grullón L, Matás C, Avilés M (2008) Oviduct-specific glycoprotein and heparin modulate sperm-zona pellucida interaction during fertilization and contribute to the control of polyspermy. Proc Natl Acad Sci U S A 105: 15809–15814

Coy P, Jimenez-Movilla M, Garcia-Vazquez FA, Mondejar I, Grullon L, Romar R (2012) Oocytes use the plasminogen-plasmin system to remove supernumerary spermatozoa. Hum Reprod 27:1985–1993

Curchoe CL, Zhang S, Yang L, Page R, Tian XC (2009) Hypomethylation trends in the intergenic region of the imprinted IGF2 and H19 genes in cloned cattle. Anim Reprod Sci 116:213–225

de Souza D, Salles L, Rosa e Silva A (2015) Aspects of energetic substrate metabolism of in vitro and in vivo bovine embryos. Braz J Med Biol Res 48:191–197

Degrelle SA, Campion E, Cabau C, Piumi F, Reinaud P, Richard C, Renard JP, Hue I (2005) Molecular evidence for a critical period in mural trophoblast development in bovine blastocysts. Dev Biol 288:448–460

Du Plessis S, Makker K, Desai N, Agarwal A (2008) Impact of oxidative stress on IVF. Expert Rev Obstet Gynecol 3:539–554

Eichenlaub-Ritter U, Peschke M (2002) Expression in in-vivo and in-vitro growing and maturing oocytes: focus on regulation of expression at the translational level. Hum Reprod Update 8:21–41

El Hajj N, Haaf T (2013) Epigenetic disturbances in in vitro cultured gametes and embryos: implications for human assisted reproduction. Fertil Steril 99:632–641

Fair T, Hyttel P, Greve T (1995) Bovine oocyte diameter in relation to maturational competence and transcriptional activity. Mol Reprod Dev 42:437–442

Farin CE, Farmer WT, Farin PW (2010) Pregnancy recognition and abnormal offspring syndrome in cattle. Reprod Fertil Dev 22:75–87

Farin PW, Piedrahita JA, Farin CE (2006) Errors in development of fetuses and placentas from in vitro-produced bovine embryos. Theriogenology 65:178–191

Fernandez-Gonzalez R, Moreira P, Bilbao A, Jimenez A, Perez-Crespo M, Ramirez MA, Rodriguez De Fonseca F, Pintado B, Gutierrez-Adan A (2004) Long-term effect of in vitro culture of mouse embryos with serum on mRNA expression of imprinting genes, development, and behavior. Proc Natl Acad Sci U S A 101:5880–5885

Forde N, Carter F, Fair T, Crowe MA, Evans AC, Spencer TE, Bazer FW, McBride R, Boland MP, O'Gaora P, Lonergan P, Roche JF (2009) Progesterone-regulated changes in endometrial gene expression contribute to advanced conceptus development in cattle. Biol Reprod 81:784–794

Gad A, Hoelker M, Besenfelder U, Havlicek V, Cinar U, Rings F, Held E, Dufort I, Sirard MA, Schellander K, Tesfaye D (2012) Molecular mechanisms and pathways involved in bovine embryonic genome activation and their regulation by alternative in vivo and in vitro culture conditions. Biol Reprod 87:100

Gardner DK, Lane M, Calderon I, Leeton J (1996) Environment of the preimplantation human embryo in vivo: metabolite analysis of oviduct and uterine fluids and metabolism of cumulus cells. Fertil Steril 65:349–353

Goncalves FS, Barretto LS, Arruda RP, Perri SH, Mingoti GZ (2010) Effect of antioxidants during bovine in vitro fertilization procedures on spermatozoa and embryo development. Reprod Domest Anim 45:129–135

Guerin P, Menezo Y (2011) Review: role of tubal environment in preimplantation embryogenesis: application to co-culture assays. Zygote 19:47–54

Gutierrez-Adan A, Lonergan P, Rizos D, Ward FA, Boland MP, Pintado B, de la Fuente J (2001) Effect of the in vitro culture system on the kinetics of blastocyst development and sex ratio of bovine embryos. Theriogenology 55:1117–1126

Hamdi M, Lopera R, Maillo V, Núñez C, Gutierrez-Adan A, Lonergan P, Bermejo-Alvarez P, Rizos D (2015) Bovine oviduct epithelial cells: an in vitro model to study early embryo-maternal communication. Anim Reprod 12:798

Hansen PJ, Block J, Loureiro B, Bonilla L, Hendricks KE (2010) Effects of gamete source and culture conditions on the competence of in vitro-produced embryos for post-transfer survival in cattle. Reprod Fertil Dev 22:59–66

Hasler JF, Henderson WB, Hurtgen PJ, Jin ZQ, McCauley AD, Mower SA, Neely B, Shuey LS, Stokes JE, Trimmer SA (1995) Production, freezing and transfer of bovine IVF embryos and subsequent calving results. Theriogenology 43:141–152

Heinzmann J, Hansmann T, Herrmann D, Wrenzycki C, Zechner U, Haaf T, Niemann H (2011) Epigenetic profile of developmentally important genes in bovine oocytes. Mol Reprod Dev 78:188–201

Hergenreider E, Heydt S, Tréguer K, Boettger T, Horrevoets AJG, Zeiher AM, Scheffer MP, Frangakis AS, Yin X, Mayr M, Braun T, Urbich C, Boon RA, Dimmeler S (2012) Atheroprotective communication between endothelial cells and smooth muscle cells through miRNAs. Nat Cell Biol 14:249–256

Holm P, Booth PJ, Schmidt MH, Greve T, Callesen H (1999) High bovine blastocyst development in a static in vitro production system using SOFaa medium supplemented with sodium citrate and myo-inositol with or without serum-proteins. Theriogenology 52:683–700

Holt WV (1997) Alternative strategies for the long-term preservation of spermatozoa. Reprod Fertil Dev 9:309–319

Hoshi H (2003) In vitro production of bovine embryos and their application for embryo transfer. Theriogenology 59:675–685

Hunter RHF (1988) The fallopian tubes: their role in fertility and infertility. Springer-Verlag, Berlin, 191 pp.

Hunter RH, Wilmut I (1984) Sperm transport in the cow: peri-ovulatory redistribution of viable cells within the oviduct. Reprod Nutr Dev 24:597–608

Jang HY, Kim YH, Kim BW, Park IC, Cheong HT, Kim JT, Park CK, Kong HS, Lee HK, Yang BK (2010) Ameliorative effects of melatonin against hydrogen peroxide-induced oxidative stress on boar sperm characteristics and subsequent in vitro embryo development. Reprod Domest Anim 45:943–950

Khosla S, Dean W, Brown D, Reik W, Feil R (2001) Culture of preimplantation mouse embryos affects fetal development and the expression of imprinted genes. Biol Reprod 64:918–926

Knijn HM, Wrenzycki C, Hendriksen PJ, Vos PL, Herrmann D, van der Weijden GC, Niemann H, Dieleman SJ (2002) Effects of oocyte maturation regimen on the relative abundance of gene transcripts in bovine blastocysts derived in vitro or in vivo. Reproduction 124:365–375

Kothari S, Thompson A, Agarwal A, du Plessis SS (2010) Free radicals: their beneficial and detrimental effects on sperm function. Indian J Exp Biol 48:425–435

Lawrence JL, Payton RR, Godkin JD, Saxton AM, Schrick FN, Edwards JL (2004) Retinol improves development of bovine oocytes compromised by heat stress during maturation. J Dairy Sci 87:2449–2454

Lazzari G, Wrenzycki C, Herrmann D, Duchi R, Kruip T, Niemann H, Galli C (2002) Cellular and molecular deviations in bovine in vitro-produced embryos are related to the large offspring syndrome. Biol Reprod 67:767–775
Livingston T, Eberhardt D, Edwards JL, Godkin J (2004) Retinol improves bovine embryonic development in vitro. Reprod Biol Endocrinol 2:83
Lloyd A, Pratt K, Siebrasse E, Moran MD, Duina AA (2009) Uncoupling of the patterns of chromatin association of different transcription elongation factors by a histone H3 mutant in Saccharomyces cerevisiae. Eukaryot Cell 8:257–260
Lonergan P (2007) State-of-the-art embryo technologies in cattle. Soc Reprod Fertil Suppl 64:315–325
Lonergan P, Fair T (2016) Maturation of oocytes in vitro. Annu Rev Anim Biosci 4:255–268
Lonergan P, Rizos D, Gutierrez-Adan A, Fair T, Boland MP (2003a) Effect of culture environment on embryo quality and gene expression – experience from animal studies. Reprod Biomed Online 7:657–663
Lonergan P, Rizos D, Kanka J, Nemcova L, Mbaye AM, Kingston M, Wade M, Duffy P, Boland MP (2003b) Temporal sensitivity of bovine embryos to culture environment after fertilization and the implications for blastocyst quality. Reproduction 126:337–346
Lopera-Vasquez R, Hamdi M, Maillo V, Lloreda V, Coy P, Gutierrez-Adan A, Bermejo-Alvarez P, Rizos D (2015) Effect of bovine oviductal fluid on development and quality of bovine embryos produced in vitro. Reprod Fertil Dev. doi:10.1071/RD15238
Lopera-Vásquez R, Hamdi M, Fernandez-Fuertes B, Maillo V, Beltrán-Breña P, Calle A, Redruello A, López-Martín S, Gutierrez-Adán A, Yañez-Mó M, Ramirez MÁ, Rizos D (2016) Extracellular vesicles from BOEC in in vitro embryo development and quality. PLoS One 11:e0148083
Maillo V, Sánchez-Calabuig MJ, Lopera-Vasquez R, Hamdi M, Gutierrez-Adan A, Lonergan P, Rizos D (2016) Oviductal response to gametes and early embryos in mammals. Reproduction 152:R127–R141
Marei WF, Ghafari F, Fouladi-Nashta AA (2012) Role of hyaluronic acid in maturation and further early embryo development of bovine oocytes. Theriogenology 78:670–677
Marques A, Santos P, Antunes G, Chaveiro A, Moreira da Silva F (2010) Effect of alpha-tocopherol on bovine in vitro fertilization. Reprod Domest Anim 45:81–85
Miessen K, Sharbati S, Einspanier R, Schoen J (2011) Modelling the porcine oviduct epithelium: a polarized in vitro system suitable for long-term cultivation. Theriogenology 76:900–910
Morris DG, Diskin MG, Sreenan JM (2000) Protein synthesis and phosphorylation by elongating 13–15-day-old cattle blastocysts. Reprod Fertil Dev 12:39–44
Murillo-Ríos A, Maillo V, Muñoz M, Gutiérrez-Adán A, Carrocera S, Martín-González D, Fernandez-Buznego A, Gómez E (2016) Short- and long-term outcomes of the absence of protein during bovine blastocyst formation in vitro. Reprod Fertil Dev 29(6):1064–1073
Ng YH, Rome S, Jalabert A, Forterre A, Singh H, Hincks CL, Salamonsen LA (2013) Endometrial exosomes/microvesicles in the uterine microenvironment: a new paradigm for embryo-endometrial cross talk at implantation. PLoS One 8:e58502
Nowak-Imialek M, Wrenzycki C, Herrmann D, Lucas-Hahn A, Lagutina I, Lemme E, Lazzari G, Galli C, Niemann H (2008) Messenger RNA expression patterns of histone-associated genes in bovine preimplantation embryos derived from different origins. Mol Reprod Dev 75:731–743
Pang YW, Sun YQ, Sun WJ, WH D, Hao HS, Zhao SJ, Zhu HB (2016) Melatonin inhibits paraquat-induced cell death in bovine preimplantation embryos. J Pineal Res 60:155–166
Paramio MT, Izquierdo D (2016) Recent advances in in vitro embryo production in small ruminants. Theriogenology 86(1):152–159
Perecin F, Meo SC, Yamazaki W, Ferreira CR, Merighe GK, Meirelles FV, Garcia JM (2009) Imprinted gene expression in in vivo- and in vitro-produced bovine embryos and chorio-allantoic membranes. Genet Mol Res 8:76–85
Perez-Cerezales S, Gutierrez-Adan A, Martinez-Paramo S, Beirao J, Herraez MP (2011) Altered gene transcription and telomere length in trout embryo and larvae obtained with DNA cryo-damaged sperm. Theriogenology 76:1234–1245

Perez-Cerezales S, Martinez-Paramo S, Beirao J, Herraez MP (2010) Fertilization capacity with rainbow trout DNA-damaged sperm and embryo developmental success. Reproduction 139:989–997

Piehl LL, Fischman ML, Hellman U, Cisale H, Miranda PV (2013) Boar seminal plasma exosomes: effect on sperm function and protein identification by sequencing. Theriogenology 79: 1071–1082

Quesenberry PJ, Aliotta J, Deregibus MC, Camussi G (2015) Role of extracellular RNA-carrying vesicles in cell differentiation and reprogramming. Stem Cell Res Ther 6:153

Raposo G, Stoorvogel W (2013) Extracellular vesicles: Exosomes, microvesicles, and friends. J Cell Biol 200:373–383

Regassa A, Rings F, Hoelker M, Cinar U, Tholen E, Looft C, Schellander K, Tesfaye D (2011) Transcriptome dynamics and molecular cross-talk between bovine oocyte and its companion cumulus cells. BMC Genomics 12:57

Reik W, Dean W, Walter J (2001) Epigenetic reprogramming in mammalian development. Science 293:1089–1093

Rizos D, Clemente M, Bermejo-Alvarez P, de La Fuente J, Lonergan P, Gutierrez-Adan A (2008) Consequences of in vitro culture conditions on embryo development and quality. Reprod Domest Anim 43(Suppl 4):44–50

Rizos D, Gutierrez-Adan A, Perez-Garnelo S, De La Fuente J, Boland MP, Lonergan P (2003) Bovine embryo culture in the presence or absence of serum: implications for blastocyst development, cryotolerance, and messenger RNA expression. Biol Reprod 68:236–243

Rizos D, Pintado B, de la Fuente J, Lonergan P, Gutiérrez-Adán A (2007) Development and pattern of mRNA relative abundance of bovine embryos cultured in the isolated mouse oviduct in organ culture. Mol Reprod Dev 74:716–723

Rizos D, Ward F, Duffy P, Boland MP, Lonergan P (2002) Consequences of bovine oocyte maturation, fertilization or early embryo development in vitro versus in vivo: implications for blastocyst yield and blastocyst quality. Mol Reprod Dev 61:234–248

Roberts RM, Ealy AD, Alexenko AP, Han CS, Ezashi T (1999) Trophoblast interferons. Placenta 20:259–264

Robinson RS, Hammond AJ, Wathes DC, Hunter MG, Mann GE (2008) Corpus luteum-endometrium-embryo interactions in the dairy cow: underlying mechanisms and clinical relevance. Reprod Domest Anim 43(Suppl 2):104–112

Roca J, Gil MA, Hernandez M, Parrilla I, Vazquez JM, Martinez EA (2004) Survival and fertility of boar spermatozoa after freeze-thawing in extender supplemented with butylated hydroxytoluene. J Androl 25:397–405

Roca J, Rodriguez MJ, Gil MA, Carvajal G, Garcia EM, Cuello C, Vazquez JM, Martinez EA (2005) Survival and in vitro fertility of boar spermatozoa frozen in the presence of superoxide dismutase and/or catalase. J Androl 26:15–24

Rottmayer R, Ulbrich SE, Kolle S, Prelle K, Neumueller C, Sinowatz F, Meyer HH, Wolf E, Hiendleder S (2006) A bovine oviduct epithelial cell suspension culture system suitable for studying embryo-maternal interactions: morphological and functional characterization. Reproduction 132:637–648

Russell DF, Baqir S, Bordignon J, Betts DH (2006) The impact of oocyte maturation media on early bovine embryonic development. Mol Reprod Dev 73:1255–1270

Saadeldin IM, Kim SJ, Choi YB, Lee BC (2014) Improvement of cloned embryos development by co-culturing with parthenotes: a possible role of exosomes/microvesicles for embryos paracrine communication. Cell Reprogram 16:223–234

Saadeldin IM, HJ O, Lee BC (2015) Embryonic-maternal cross-talk via exosomes: potential implications. Stem Cells Cloning 8:103–107

Sapanidou V, Taitzoglou I, Tsakmakidis I, Kourtzelis I, Fletouris D, Theodoridis A, Zervos I, Tsantarliotou M (2015) Antioxidant effect of crocin on bovine sperm quality and in vitro fertilization. Theriogenology 84:1273–1282

Senger P (2003) Pathways to pregnancy and parturition. Current conceptions, Inc., Pullman

Silva PF, Gadella BM, Colenbrander B, Roelen BA (2007) Exposure of bovine sperm to pro-oxidants impairs the developmental competence of the embryo after the first cleavage. Theriogenology 67:609–619

Silveira JC, Veeramachaneni DNR, Winger QA, Carnevale EM, Bouma GJ (2012) Cell-secreted vesicles in equine ovarian follicular fluid contain miRNAs and proteins: a possible new form of cell communication within the ovarian follicle. Biol Reprod 86:71

Simons M, Raposo G (2009) Exosomes – vesicular carriers for intercellular communication. Curr Opin Cell Biol 21:575–581

Simpson RJ, Jensen SS, Lim JWE (2008) Proteomic profiling of exosomes: current perspectives. Proteomics 8:4083–4099

Sirard MA (2001) Resumption of meiosis: mechanism involved in meiotic progression and its relation with developmental competence. Theriogenology 55:1241–1254

Sirard MA, Richard F, Blondin P, Robert C (2006) Contribution of the oocyte to embryo quality. Theriogenology 65:126–136

Somfai T, Ozawa M, Noguchi J, Kaneko H, Kuriani Karja NW, Farhudin M, Dinnyes A, Nagai T, Kikuchi K (2007) Developmental competence of in vitro-fertilized porcine oocytes after in vitro maturation and solid surface vitrification: effect of cryopreservation on oocyte antioxidative system and cell cycle stage. Cryobiology 55:115–126

Spencer TE, Bazer FW (2004) Conceptus signals for establishment and maintenance of pregnancy. Reprod Biol Endocrinol 2:49

Spencer TE, Forde N, Lonergan P (2016) The role of progesterone and conceptus-derived factors in uterine biology during early pregnancy in ruminants. J Dairy Sci 99:5941–5950

Suarez SS (2008) Regulation of sperm storage and movement in the mammalian oviduct. Int J Dev Biol 52:455–462

Suzuki J Jr, Therrien J, Filion F, Lefebvre R, Goff AK, Smith LC (2009) In vitro culture and somatic cell nuclear transfer affect imprinting of SNRPN gene in pre- and post-implantation stages of development in cattle. BMC Dev Biol 9:9

Takahashi M (2012) Oxidative stress and redox regulation on in vitro development of mammalian embryos. J Reprod Dev 58:1–9

Tarin JJ, Perez-Albala S, Cano A (2000) Consequences on offspring of abnormal function in ageing gametes. Hum Reprod Update 6:532–549

Teijeiro JM, Marini PE (2012) The effect of oviductal deleted in malignant brain tumor 1 over porcine sperm is mediated by a signal transduction pathway that involves pro-AKAP4 phosphorylation. Reproduction 143:773–785

Tervit HR, Whittingham DG, Rowson LE (1972) Successful culture in vitro of sheep and cattle ova. J Reprod Fertil 30:493–497

Tesfaye D, Ghanem N, Carter F, Fair T, Sirard MA, Hoelker M, Schellander K, Lonergan P (2009) Gene expression profile of cumulus cells derived from cumulus-oocyte complexes matured either in vivo or in vitro. Reprod Fertil Dev 21:451–461

Tesfaye D, Lonergan P, Hoelker M, Rings F, Nganvongpanit K, Havlicek V, Besenfelder U, Jennen D, Tholen E, Schellander K (2007) Suppression of connexin 43 and E-cadherin transcripts in in vitro derived bovine embryos following culture in vitro or in vivo in the homologous bovine oviduct. Mol Reprod Dev 74:978–988

Thibodeaux JK, Roussel JD, White KL, Broussard JR, Godke RA (1992) The use of image analysis to evaluate the development of uterine and oviduct epithelial cells during in vitro culture. A potential quality assurance procedure for in vitro laboratories. Arch Pathol Lab Med 116:444–448

Thompson JG (2000) In vitro culture and embryo metabolism of cattle and sheep embryos – a decade of achievement. Anim Reprod Sci 60-61:263–275

Ulbrich SE, Zitta K, Hiendleder S, Wolf E (2010) In vitro systems for intercepting early embryo-maternal cross-talk in the bovine oviduct. Theriogenology 73:802–816

Urrego R, Rodriguez-Osorio N, Niemann H (2014) Epigenetic disorders and altered gene expression after use of assisted reproductive technologies in domestic cattle. Epigenetics 9:803–815

Van Hoeck V, Leroy JL, Arias Alvarez M, Rizos D, Gutierrez-Adan A, Schnorbusch K, Bols PE, Leese HJ, Sturmey RG (2013) Oocyte developmental failure in response to elevated nonesterified fatty acid concentrations: mechanistic insights. Reproduction 145:33–44

Van Hoeck V, Sturmey RG, Bermejo-Alvarez P, Rizos D, Gutierrez-Adan A, Leese HJ, Bols PE, Leroy JL (2011) Elevated non-esterified fatty acid concentrations during bovine oocyte maturation compromise early embryo physiology. PLoS One 6:e23183

Walker SK, Hartwich K, Seamark R (1996) The production of unusually large offspring following embryo manipulation: concepts and challenges. Theriogenology 45:111–120

Walter I (1995) Culture of bovine oviduct epithelial cells (BOEC). Anat Rec 243:347–356

Wang H, Dey SK (2006) Roadmap to embryo implantation: clues from mouse models. Nat Rev Genet 7:185–199

Wang X, Falcone T, Attaran M, Goldberg JM, Agarwal A, Sharma RK (2002) Vitamin C and vitamin E supplementation reduce oxidative stress-induced embryo toxicity and improve the blastocyst development rate. Fertil Steril 78:1272–1277

Watson PF (2000) The causes of reduced fertility with cryopreserved semen. Anim Reprod Sci 60-61:481–492

Wheeler MB, Rutledge JJ, Fischer-Brown A, VanEtten T, Malusky S, Beebe DJ (2006) Application of sexed semen technology to in vitro embryo production in cattle. Theriogenology 65:219–227

Wilson RD, Fricke PM, Leibfried-Rutledge ML, Rutledge JJ, Penfield CM, Weigel KA (2006) In vitro production of bovine embryos using sex-sorted sperm. Theriogenology 65:1007–1015

Wrenzycki C, Herrmann D, Lucas-Hahn A, Korsawe K, Lemme E, Niemann H (2005) Messenger RNA expression patterns in bovine embryos derived from in vitro procedures and their implications for development. Reprod Fertil Dev 17:23–35

Wrenzycki C, Herrmann D, Niemann H (2007) Messenger RNA in oocytes and embryos in relation to embryo viability. Theriogenology 68(Suppl 1):S77–S83

Young LE, Fernandes K, McEvoy TG, Butterwith SC, Gutierrez CG, Carolan C, Broadbent PJ, Robinson JJ, Wilmut I, Sinclair KD (2001) Epigenetic change in IGF2R is associated with fetal overgrowth after sheep embryo culture. Nat Genet 27:153–154

Zullo G, De Canditiis C, Pero ME, Albero G, Salzano A, Neglia G, Campanile G, Gasparrini B (2016) Crocetin improves the quality of in vitro-produced bovine embryos: implications for blastocyst development, cryotolerance, and apoptosis. Theriogenology 86(8):1879–1885

The Role of Maternal Nutrition During the Periconceptional Period and Its Effect on Offspring Phenotype

Tom P. Fleming, Judith J. Eckert, and Oleg Denisenko

Abstract The early preimplantation embryo has been rigorously studied for decades to understand inherent reproductive and developmental mechanisms driving its morphogenesis from before fertilisation through to and beyond implantation. Recent research has demonstrated that this short developmental window is also critical for the embryo's interaction with external, maternal factors, particularly nutritional status. Here, maternal dietary quality has been shown to alter the pattern of development in an enduring way that can influence health throughout the lifetime. Thus, using mouse models, maternal protein restriction exclusively during the preimplantation period with normal nutrition thereafter is sufficient to cause adverse cardiometabolic and neurological outcomes in adult offspring. Evidence for similar effects whereby environmental factors during the periconceptional window can programme postnatal disease risk can be found in human and large animal models and also in response to in vitro conditions such as assisted conception and related infertility treatments. In this review, using mouse malnutrition models, we evaluate the step-by-step mechanisms that lead from maternal poor diet consumption though to offspring disease. We consider how adverse programming within the embryo may be induced, what nutrient factors and signalling pathways may be involved, and how these cues act to change the embryo in distinct ways across placental and foetal lineage paths, leading especially to changes in the growth trajectory which in turn associate with later disease risk. These mechanisms straddle epigenetic, molecular, cellular and physiological levels of biology and suggest, for health outcomes, preimplantation development to be the most important time in our lives.

T.P. Fleming (✉)
Biological Sciences, University of Southampton, Mailpoint 840, Southampton General Hospital, Southampton SO16 6YD, UK
e-mail: tpf@soton.ac.uk

J.J. Eckert
Faculty of Medicine, University of Southampton, Mailpoint 840, Southampton General Hospital, Southampton SO16 6YD, UK

O. Denisenko
Department of Medicine, University of Washington,
850 Republican St, Seattle, WA 98109, USA

A. Fazeli, W.V. Holt (eds.), *Periconception in Physiology and Medicine*,
Advances in Experimental Medicine and Biology 1014,
DOI 10.1007/978-3-319-62414-3_5

Keywords Preimplantation embryo • Blastocyst • Maternal nutrition • Low protein diet • Developmental programming • Growth trajectory • Extra-embryonic lineages • Ribosome biogenesis • Amino acids • Endocytosis

1 Introduction

Manipulation of reproduction during the periconceptional period is now commonplace across mammalian species. Domestic farm animals are bred to maximise milk or meat production and quality through in vitro oocyte maturation, in vitro fertilisation (IVF), embryo culture and transfer together with other contributing biotechnologies such as cloning and cryopreservation (Li et al. 2013). Similarly, overcoming human infertility, or the passage of adverse genetic risk to children, has facilitated related reproductive treatments (but not cloning!) to be developed clinically. Such assisted reproductive treatments (ART) have been largely successful in promoting efficient food production on one hand and safe delivery of several million babies globally on the other (Nardelli et al. 2014). Whilst development of such *direct* in vitro reproductive technologies has flourished, our understanding of more subtle in vivo influences on the periconception gametes and embryos has also increased, largely with the same goal, to improve reproductive outcome and health in the next generation. Thus, sustained, high quality maternal nutrition is important in enhancing reproductive performance in domestic animals (Diskin and Kenny 2016) and provision of such nutrition close to the time of conception has been recognised (Fleming et al. 2012; Leroy et al. 2015). Overcoming human metabolic disorders such as obesity and diabetes by healthy diet before conception is now advised to improve fertility and reproductive outcomes (Barker et al. 2013; Dodd et al. 2015). The same is true for women undergoing ART treatment to increase fertility success (Sim et al. 2014).

Periconceptional reproductive manipulation therefore provides an opportunity to overcome some of the obstacles we face in food security, clinical infertility and the wish for healthy offspring. But such optimism must be checked; recent research suggests that environment and manipulations around the period of conception can also cause *adverse* long-term outcomes, the opposite of the desired goal. In fact, the periconceptional period is a complex stage that can act as a 'double-edged sword' in terms of desired outcomes. One of the first inklings of this came following the early manipulation of sheep embryos in vitro and the unexpected consequence of 'large offspring syndrome' – the occurrence of abnormal excess growth at foetal and neonatal stages after extended culture of preimplantation embryos before transfer, commonly a lethal condition (McEvoy et al. 2000). Close scrutiny of the health of IVF children has revealed a small but increased incidence of perinatal complications (such as growth restriction and prematurity), imprinted gene disorders, and congenital abnormalities (Hart and Norman 2013; Brison et al. 2013; Lazaraviciute et al. 2014). For in vivo maternal nutrition and physiology, the pioneering epidemiological studies from David Barker and colleagues (Southampton) on what has become known as the Developmental Origins of Health and Disease (DOHaD) hypothesis has demonstrated poor maternal nutrition during pregnancy closely associates with increased risks of adult non-communicable cardiometabolic disease in offspring (Barker 2007; Barker and Thornburg 2013). These studies across human clinical and animal models have further shown that poor maternal nutrition particularly during the periconceptional period in vivo can programme a broad spectrum of adult offspring comorbidities associated with cardiovascular, metabolic and neurological disease (de Rooij and Roseboom 2013; Turner and Robker 2015; Fleming et al. 2015). Moreover, human ART studies indicate increased prevalence of cardiometabolic disease risk in IVF children compared with spontaneously conceived children from parents with infertility problems (Ceelen et al. 2008).

The focus of attention on periconceptional reproduction, both from in vitro ART and in vivo maternal nutrition and physiology, leads us to consider two related questions. *Why might the period around conception be so susceptible to manipulations that it can alter long-term health outcomes throughout the lifespan and across species?* This question concerns the biological characteristics of the egg and early embryo, how the events of early development, both molecular and morphogenetic, may be instrumental in permitting manipulative conditions to impose immediate and enduring change on the developmental programme. In this review, we first briefly discuss the biological characteristics of the periconceptional period with these issues in mind. The second question to consider is – *How might environmental conditions experienced by the egg and early embryo change long-term potential affecting health across the lifespan?* This question concerns firstly the interface between the egg/embryo and different external factors, the sensing mechanisms that may 'read' this interface, and then how these conditions may cause permanent change in the way the embryo develops. The review will secondly discuss this multistep process using mainly a rodent maternal undernutrition model that has revealed some understanding of the plethora of mechanisms at work here.

2 Early Embryo Biology and Its Susceptibility to Developmental Programming

From fertilisation, the embryo has a multitude of tasks to achieve within a brief period of a few days. The highly specialised gametes must 'de-differentiate' and transform into a totipotent zygote and reinitiate cell cycling (cleavage divisions) coupled with cessation of expression of the parental genomes and activation of the new and unique embryonic genome (Li et al. 2013). Early blastomeres must begin the processes of expression and post-translational modifications regulating intercellular adhesion at compaction (8-cell stage in mouse) and subsequent epithelial differentiation in outer cells forming the extra-embryonic trophectoderm of the blastocyst (32-cell stage in mouse) (Eckert and Fleming 2008; Eckert et al. 2015).

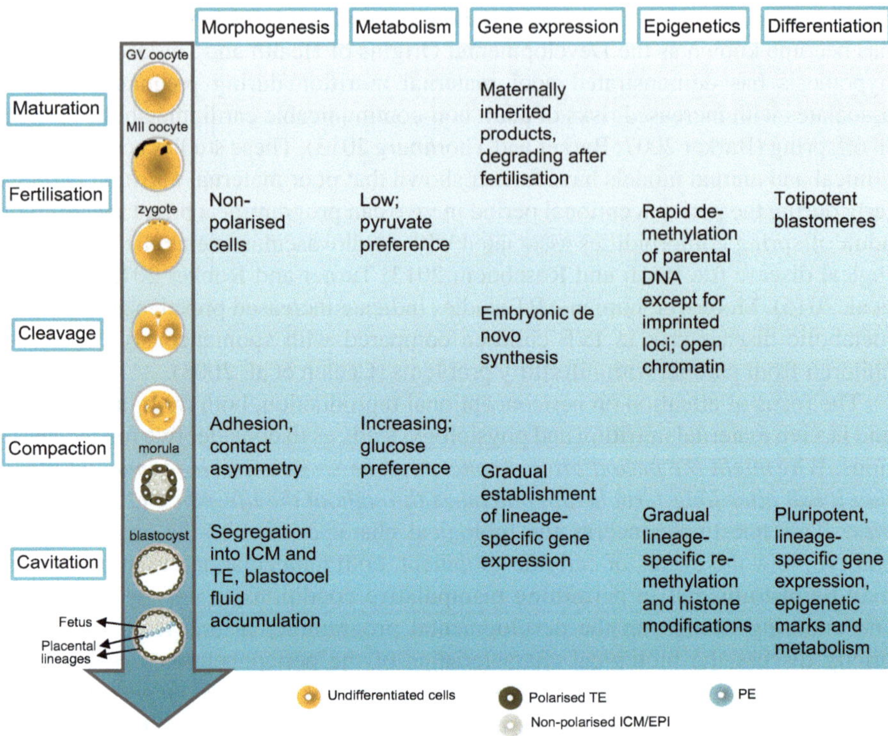

Fig. 1 A summary of the multiple events underway simultaneously in the early embryo activated upon maturation of the oocyte and its fertilisation. The proximity of these intrinsic events make the early embryo a suitable developmental stage for 'sensing' environmental conditions and optimising its future development (plasticity) dependent upon these conditions, including the quality of maternal nutrition. For details and references, see the text

Concomitant with embryo cleavage is the diversification and segregation of embryonic and extra-embryonic cell lineages (forming inner cell mass (ICM) internal to trophectoderm), achieved essentially through coordinated asymmetric cell divisions of polarised cells and through differential expression of regulatory transcription factors (Cockburn and Rossant 2010; Bedzhov et al. 2014; Sozen et al. 2014). Embryo morphogenesis further includes maturation of metabolic and nutrient uptake pathways (Leese 2012; Gardner and Harvey 2015), Trophectoderm transepithelial transport activity to form the blastocoel cavity (Eckert and Fleming 2008; Chen et al. 2010), and later hatching from the zona pellucida and the coordinated attachment to the uterine endometrial surface to achieve implantation (Seshagiri et al. 2009; Cha et al. 2012). See Fig. 1 for a summary of these events.

This complex series of biological processes, activated by fertilisation and rapidly accomplished, is essential for continuation of the developmental programme. The switch from gamete to embryo stages and the emergence of distinct cell lineages requires a major epigenetic reorganisation of the genome. This comprises an initial global DNA demethylation of paternal and maternal genomes in blastomeres whilst maintaining the allele-specific pattern of imprinted gene expression and, from the blastocyst stage, the gradual re-methylation of genes that is lineage-specific and essential for their explicit patterns of gene expression (Chen et al. 2010; Rivera and Ross 2013; Zhou and Dean 2015).

Early embryo biology therefore comprises the coupling of several molecular, genetic and cellular activities over a strict time course and must coordinate the completion of maturation with maternal endometrial receptivity for development to continue. Thus, small changes or aberrations in any of these formative events may act to block or perturb the progression of a healthy pregnancy. However, we believe this is only half the story! In addition to the *intrinsic* series of biological processes governing the onset of gestation, the periconceptional period offers a unique 'opportunity' for *developmental plasticity*, a process to optimise developmental phenotype from a single genotype according to *extrinsic* conditions experienced by the embryo. The concept of embryo plasticity means that diverse environmental conditions can be 'factored' into the developmental programme as a conserved mechanism to promote offspring survival and competitiveness. Environmental conditions including maternal nutrition, metabolic health, sickness and infection could therefore act via appropriate embryo 'sensing' mechanisms to modify or adapt the intrinsic developmental programme. The epigenetic restructuring of the genome in the early embryo (Chen et al. 2010; Rivera and Ross 2013; Zhou and Dean 2015) provides a way for environmental sensing signals to induce modifications in gene expression and for these to persist in appropriate cell lineages long after the sensing event has occurred. In the context of maternal nutrition and its effect on development, we have found such adaptations can be distinct between embryonic and extra-embryonic cell lineages during the periconceptional period, to be mediated through epigenetic changes, and to confer an advantage in offspring fitness (Sun et al. 2015; Denisenko et al. 2016). We discuss such examples below. However, developmental plasticity mediates advantage usually only if environmental conditions persist and can represent the 'double-edged sword' referred to above in the Introduction when conditions change and adaptations no longer match conditions experienced in utero. Here, the long-term consequences can be increased susceptibility to disease comorbidities in offspring and can occur not only from in vivo conditions but also those experienced in vitro through ART manipulations.

3 Maternal Periconceptional Nutrition and Embryo Developmental Plasticity and Potential

The major animal studies on maternal periconceptional nutrition have utilised the rodent low protein diet (LPD) model where protein restriction (9% casein vs control 18%, normal protein diet (NPD)) is limited exclusively to the preimplantation period (E0, 3.5 in mouse; E0, 4.25 in rat) with normal healthy nutrition thereafter for gestation and throughout postnatal life (known as Emb-LPD treatment). The maternal Emb-LPD diet leads to adult offspring comorbidity of abnormal cardiovascular, metabolic, growth and behavioural phenotype (Watkins et al. 2008a, 2010, 2011; Kwong et al. 2000). This profound legacy from a single dietary challenge mediated to their mothers when they were cleavage stage embryos is a powerful demonstration of adverse developmental programming that we find in this model is more acutely sensitive in female offspring. In more detail, outcomes include hypertension throughout life accompanied by attenuated arterial capacity to dilate and increased angiotensin converting enzyme activity; high adiposity and acquisition of a storage-type metabolism adapted to retain fats; increased growth and weight throughout life; and an abnormal hyperactive behaviour pattern (Watkins et al. 2008a, 2010, 2011). A similar yet distinct adult offspring phenotype can occur if the maternal LPD window is targeted to the period of oocyte maturation before fertilisation rather than of the early embryo (Watkins et al. 2008b), suggesting the periconceptional window extends before conception but that the presence of the paternal genome is contributory to the full phenotype, further supported by administration of paternal LPD alone (Watkins and Sinclair 2014). The influence and legacy of poor maternal nutrition during the periconceptional period has also been observed in domestic mammals using related models. Thus, ewes fed an undernutrition diet during the periconceptional period give rise to offspring with increased risk of cardiovascular, metabolic and neurocognitive disorders (Gardner et al. 2004; Hernandez et al. 2009; Todd et al. 2009; Torrens et al. 2009).

There is also good evidence that similar consequences occur in the human. People *conceived* during the 5-month Dutch Hunger Winter during the Second World War had increased risk of coronary heart disease, hypertension, high BMI, glucose intolerance, schizophrenia and depression and poorer cognition in adulthood compared to those experiencing the famine *later* in pregnancy (Painter et al. 2006; Roseboom et al. 2011; de Rooij and Roseboom 2013). Similarly, maternal nutritional profile at the time of conception in women in The Gambia has been associated with lifelong health of offspring born (Moore et al. 1997). In both of these human models, epigenetic modifications in a number of genes have been identified in offspring that associate with the periconceptional environmental challenge mediated through maternal nutrition (Heijmans et al. 2008; Waterland et al. 2010; Dominguez-Salas et al. 2014). Below, we will consider the stepwise mechanisms including epigenetic processes linking maternal periconceptional nutrition with long-term health outcomes, focusing mainly on the rodent Emb-LPD model.

4 Stepwise Mechanisms in Emb-LPD Periconceptional Developmental Programming

4.1 From Maternal Environment to Embryonic Induction

To understand why preimplantation embryo nutritional programming may lead to disease in later life, we need to consider the sequence of events that we think take place. The diet of the mother first modifies metabolite composition within her circulatory system, resulting in changes evident within serum. In response to short duration Emb-LPD in mice and rats, serum amino acids and insulin levels reduce within the period of preimplantation development whilst glucose increases in concentration (Kwong et al. 2000; Eckert et al. 2012). For the embryo to 'perceive' the maternal nutritional condition, such changes must be transmitted into the female reproductive tract where the embryos are located in the lumen before implantation. Careful collection and biochemical analysis of the uterine tract fluid from Emb-LPD and NPD dams at the time of blastocyst formation has shown that, at least for amino acids, the reduction in their concentration after Emb-LPD is also evident here (Eckert et al. 2012), indicating maternal dietary activity can modify tract secretions. A similar dietary influence on human uterine fluid amino acid composition has been shown recently (Kermack et al. 2015). Given the close juxtaposition of altered tract amino acid composition following Emb-LPD and the blastocyst, we must consider what signalling mechanisms might transmit this information to the embryo to initiate developmental programming. The mammalian target of rapamycin complex 1 (mTORC1) signalling pathway is the well-recognised regulator of cellular growth and translation rate, and its activity is mediated through extracellular insulin (acting through the insulin receptor and downstream phosphoinositide 3-kinase, PI3K) together with select amino acids of the branched chain group (BCAAs; leucine, valine, isoleucine) that contribute to cellular signalling (Avruch et al. 2009; Dibble and Cantley 2015). Indeed, BCAAs, both collectively and individually, are depleted in the serum and uterine fluid following Emb-LPD and changes in several amino acids have also been recorded in Emb-LPD blastocysts (Eckert et al. 2012). Moreover, quantitative analysis of mTORC1 in Emb-LPD blastocysts has revealed a significant reduction in activity of the downstream effector S6 ribosomal protein, a translation activator and regulator (Kim 2009). Collectively, these data suggest that maternal dietary Emb-LPD can alter nutrient composition within the uterus and that blastocysts can recognise this through sensing mechanisms that reduce mTORC1 signalling, representing the inductive event underlying developmental programming.

4.2 Programming of the Growth Trajectory

Following receipt of the reduced maternal nutrient environment, the blastocyst undergoes a series of responses or adaptations in the way it develops that act to compensate for the maternal nutrient restriction. These changes (discussed in 5.3–5.4) are distinct between the embryonic and extra-embryonic lineages, occur at different times during gestation, but act co-ordinately to promote survival and competitiveness. Perhaps most critical with respect to a mechanistic understanding of the long-term outcomes affecting growth, cardiovascular, metabolic and behavioural consequences are those responses that influence the growth trajectory of the embryo and foetus. We have found that the weight of pups around birth from Emb-LPD (or LPD, when the nutrient challenge is administered throughout gestation) is positively correlated with offspring comorbidity risks of adult weight, cardiovascular and behavioural phenotypes (Watkins et al. 2008a). The increase in weight caused by Emb-LPD treatment is evident at birth and, in females, is maintained relative to NPD offspring throughout life. This weight advantage is also evident during earlier foetal life and associates with an increased placental efficiency (foetal-placental weight ratio) in both Emb-LPD and LPD foetuses at E17.5 (Watkins et al. 2008a, 2015). This prenatal response reflects different adaptations occurring within embryonic and placental lineages activated from the blastocyst stage (see below). In LPD foetuses (untested in Emb-LPD foetuses), placental-foetal nutrient transfer activity is increased during mid/late gestational periods (Coan et al. 2011). The changes in blastocyst developmental path triggered by Emb-LPD become programmed at this time point because transfer of Emb-LPD and NPD blastocysts into NPD foster mothers results in enhanced growth of the Emb-LPD conceptuses in late gestation even in the absence of a continued Emb-LPD maternal background (Watkins et al. 2008a). Thus, we consider the Emb-LPD amino acid and metabolite induction mechanism involving mTORC1 signalling causes programming of blastocysts which initiates the early responses found within embryonic and extra-embryonic lineages to protect gestational growth.

4.3 Extra-Embryonic Compensatory Responses

The blastocyst outer trophectoderm layer shows several cell biological changes in response to maternal Emb-LPD that collectively contribute to the stimulation in growth discussed above. First, the Emb-LPD trophectoderm displays an increase in cellular endocytosis that likely reflects a response to increase nutrient uptake quantitatively from the maternal uterine fluid using histiotrophic means. This response appears to be comprehensive in that fluid uptake, protein ligand uptake, lysosome production and endocytosis receptor (megalin) expression are all significantly enhanced although no effect on cellular autophagy is evident (Sun et al. 2014). Mechanistically, trophectoderm endocytosis stimulation is brought about by reorganisation of the actin cytoskeleton, mediated at least in part by Rho-A GTPase activity. In vitro studies suggest the trophectoderm endocytic response to maternal protein restriction is activated by the depleted uterine amino acid environment that the diet causes and is brought about rapidly and irreversibly in a culture medium with low concentration of protein (Sun et al. 2014). The mouse trophectoderm also undergoes a slight increase in cellular proliferation in response to Emb-LPD although this is not the case in the rat trophectoderm (Eckert et al. 2012; Kwong et al. 2000). Lastly, the cellular motility that accompanies the transition of late blastocyst trophectoderm (after hatching and attachment) into migratory trophoblast cells is significantly increased by maternal Emb-LPD. This response has been demonstrated in vitro in a normalised culture environment to mimic the implantation process, several days after blastocyst collection from mothers, indicating that its induction does not require continued protein restriction (Eckert et al. 2012). Indeed, programmed changes in Emb-LPD and especially LPD trophoblast motility persist in vivo and can be detected by in vitro assays at E8.5 (Watkins et al. 2015).

The primitive endoderm extra-embryonic lineage forms on the blastocoelic face of the ICM in the late blastocyst and gives rise to parietal and visceral endoderm, the latter contributing to the histiotrophic uterine-foetal nutrient transfer activity of the yolk sac placenta (Beckman et al. 1997; Zohn and Sarkar 2010). Examination of the mouse primitive endoderm phenotype in response to Emb-LPD has been conducted using embryonic stem cell lines (ESCs) generated from blastocysts collected from diet-treated mothers. Here, primitive endoderm in embryoid bodies grown in vitro from Emb-LPD ESCs show the same increased endocytosis phenotype as described above for the trophectoderm (Sun et al. 2014). This phenotype is also evident in vivo in the isolated yolk sac collected from E17.5 LPD conceptuses (Watkins et al. 2008a).

A common compensatory strategy in both trophectoderm and primitive endoderm lineages in response to maternal protein restriction is, therefore, to increase histiotrophic nutrition through endocytosis. The capacity to ‘programme’ these responses from preimplantation environment through to late gestation, likely requires epigenetic modifications as discussed above. With respect to Emb-LPD induced changes in primitive endoderm, we have found epigenetic modifications to occur within the gene encoding the Gata6 transcription factor that is known to activate primitive endoderm specification and differentiation through several downstream target genes (Rossant et al. 2003; Schrode et al. 2014; Artus et al. 2011; Morrisey et al. 2000). We noted that the embryoid bodies with outer primitive endoderm-like layer derived from Emb-LPD ESCs grew in normalised culture conditions to a significantly increased size (~15% larger) compared with NPD embryoid bodies yet exhibit reduced expression of Gata6 at mRNA and protein levels, as occurs in the in vivo Emb-LPD yolk sac visceral endoderm (Sun et al. 2015). Reduced Gata6 expression has been observed in other cellular systems regulated by this transcription factor such as in ovarian cancer models and coincides with increased growth promotion and malignancy (Caslini et al. 2006; Capo-Chichi et al. 2010; Cai et al. 2009). Increased proliferation of primitive endoderm and suppression of differentiation in response to Emb-LPD may facilitate histiotrophic capacity in later gestation. Upon examination of the *Gata6* promoter in Emb-LPD ESCs, the reduced expression of the *Gata6* gene coincided with histone H3 and H4 hypoacetylation and reduced RNA polymerase II recruitment (Sun et al. 2015), features expected to cause reduced gene expression (Bartova et al. 2008). These epigenetic changes were coupled with increased expression of the histone deacetylase, *Hdac-1*, in Emb-LPD embryoid bodies. Interestingly, Hdac-1 expression in early embryos is sensitive to culture environment (Liu et al. 2014) and so may be a good candidate to reorganise targeted histone epigenetic reprogramming.

4.4 Embryonic Compensatory Responses

We anticipate the distinct extra-embryonic compensatory modifications discussed above, initiated in response to maternal protein restriction and mediated at least in part by epigenetic modifications, will act to enhance nutrient delivery from mother to embryo/foetus during gestation and minimise the effect of the dietary challenge. We have further investigated whether other modifications may occur within embryonic cell lineages that, mechanistically, may promote environmentally mediated control of offspring growth and thereby survival. Earlier human studies within the DOHaD literature recognised the close association that existed between foetal and postnatal growth and nutrient availability such that common characteristics for adversely programmed offspring following maternal undernutrition were low birth weight and excessive postnatal 'catch-up' growth when nutrients became unrestricted and therefore a mismatch with in utero conditions (Barker 2007; Barker and Thornburg 2013). This 'patterning' of the growth trajectory led to the 'thrifty phenotype' hypothesis which proposed poor maternal conditions would promote foetal adaptations to restrict growth of somatic tissues, modify metabolic homeostasis to conserve energy for essential organs such as the heart and brain, yet be permissive for opportunistic nutrient storage if conditions improved. However, these responses would become maladaptive if nutrient levels were significantly enriched postnatally leading to excess growth and adiposity and adult cardiometabolic disease risk (Hales and Barker 2001; Gluckman et al. 2005). This concept could apply to the Emb-LPD model since growth promotion and a storage metabolism are evident after release from the maternal nutrient challenge (Watkins et al. 2008a, 2011). In this case, however, the 'catch-up' growth is activated after implantation rather than postnatally, mediated at least in part by the compensatory responses from the extra-embryonic lineages discussed above. This demonstrates the importance of the periconceptional period for initiating long-term programming. However, despite an expansive literature associating the growth trajectory with nutrient availability and the relevance of a thrifty metabolism, clear mechanisms at genetic and epigenetic levels to explain this phenomenon have been underexplored.

In pig and sheep models we have found that protein restriction (pig) and nutrient/oxygen deprivation (sheep) triggered global downregulation of transcription in foetal kidneys (Denisenko et al. 2011). Recently, we have investigated whether a unifying mechanism might directly regulate cellular growth across different somatic tissues in response to nutrient provision using the mouse Emb-LPD and LPD model. We have found foetal and postnatal liver and kidney global transcriptional activity per cell (as reflected by cellular RNA content) to be intimately linked with nutrient availability. Thus, maternal LPD throughout gestation lead to reduced transcriptional activity in foetal somatic tissues compared with NPD controls whilst in the adult, after release from nutrient challenge, expression was increased above control levels. In the Emb-LPD tissues, both foetal and postnatal time points revealed elevated transcriptional activity beyond the NPD control (Denisenko et al. 2016). Thus, transcriptional activity per cell is linked to maternal nutrient availability with activity suppressed during nutrient challenge and stimulated after release from challenge relative to NPD controls. Looking for molecular mechanisms that regulate offspring growth in response to changes in maternal diet, we examined expression of ribosomal RNA, the major constituent of cellular RNA, and found rDNA transcriptional rate (estimated by measuring pre-rRNA levels and RNA Pol I binding to rDNA) to match the pattern of total RNA expression, implicating ribosome biogenesis as a key regulator of growth, coordinated by maternal nutrition. Epigenetic analysis of the rDNA gene promoter in liver and kidney revealed increased DNA methylation relative to controls during periods of nutrient challenge (LPD, foetal tissues) and the converse, reduced DNA methylation and increased RNA Pol1 binding after release from challenge (Emb-LPD foetal and adult tissues; LPD adult tissues). Moreover, we find some evidence that cell sensing of nutrient availability in this mechanism may be controlled through the ribosomal transcription factor, Rrn3, whose cellular levels are low during nutrient deprivation and high after release (Denisenko et al. 2016). Thus, for example, over-expression of Rrn3 in an in vitro assay stimulates cellular rDNA transcription and reduces rDNA methylation. These findings provide strong evidence that somatic tissue growth throughout the life-course can be directly regulated by nutrient availability in accordance with the thrifty phenotype hypothesis mediated through epigenetic modification of rDNA genes and the dynamics of ribosome biogenesis. We speculate, therefore, that the pattern of rDNA expression during nutritional challenge and after release fulfils the requirements of a 'thrifty gene' (Denisenko et al. 2016).

4.5 Integration of Different Mechanisms and Their Consequences

The Emb-LPD model demonstrates that an important interaction occurs between the early embryo and its maternal environment. Thus, dietary nutrient levels of the mother change the chemical milieu of the uterine fluids and embryos can sense these changes via signalling pathways. Sensing dietary nutrient levels allows the embryo to exercise some plasticity with regard to optimal phenotypic state, in particular affecting the efficiency of extra-embryonic lineages for nutrient delivery during gestation and the metabolic and biosynthetic features of the embryo. These changes appear compensatory and targeted to select a growth trajectory that is both sustainable and best fits the conditions experienced. See Fig. 2 for a summary of these integrated events.

Developmental programming following the preimplantation interaction between mother and embryo is broadly an *enduring* response, persisting throughout gestation and beyond, yet some features retain a dynamic character, and are modified by changing conditions. We demonstrate that epigenetic changes may contribute to the persistence of the programming response, for example either to key transcription factors regulating lineage diversification such as *Gata6* in the PE or to core genes

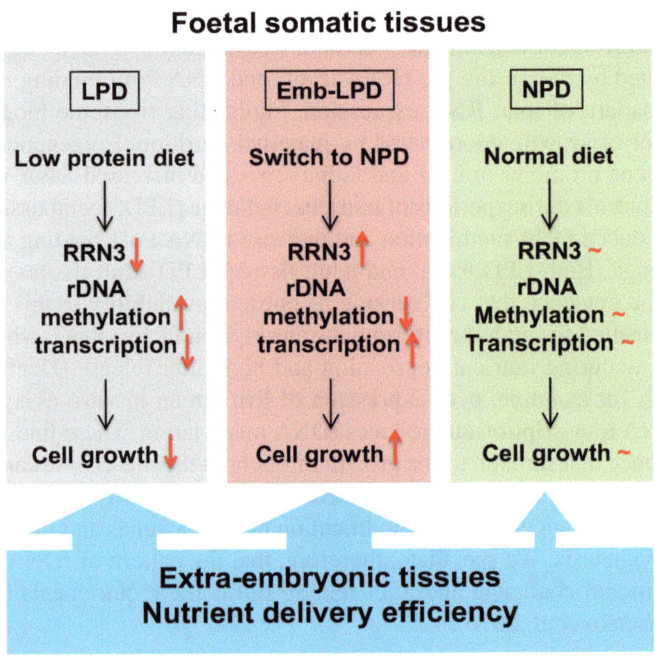

Fig. 2 The effect of maternal protein restriction on developmental plasticity in the early embryo is distinct between embryonic and extra-embryonic lineages and collectively sets the growth trajectory of the foetus. Foetal somatic tissues (e.g. liver, kidney) respond to LPD by epigenetic modification of ribosome biogenesis to restrict growth during nutrient deprivation relative to control (*NPD*). However, after release from maternal nutrient challenge (*Emb-LPD*), ribosome biogenesis is upregulated to stimulate foetal growth beyond that of NPD. In contrast, the extra-embryonic lineages (progenitors of chorioallantoic and yolk sac placentas) of the early embryo undergo several adaptations in response to maternal protein restriction that collectively promote nutrient delivery from the mother throughout gestation. These adaptations act as compensatory responses to promote growth and protect offspring fitness during dietary alterations and may involve epigenetic regulation, and most are maintained even if nutrient challenge is lifted (*Emb-LPD*). The combination of these embryonic and extra-embryonic mechanisms activated in the preimplantation embryo influence organism growth and disease risk over a lifetime. *Red arrows* represent either increased or decreased expression/activity relative to NPD control (~)

involved in biosynthesis such as rDNA in somatic tissues. The persistence of early embryonic programming after the decisive preimplantation period is most evident in the extra-embryonic lineages where compensatory responses are maintained in later gestation even if the dietary challenge is reversed (Emb-LPD) and responses become unnecessary, leading to larger offspring by birth. However, continued environmental interactions must persist to fine-tune the programming response. Thus, the increase in secondary trophoblast migratory activity at E8.5 induced by Emb-LPD is more substantial if the dietary challenge is maintained beyond implantation (LPD) (Watkins et al. 2015). Similarly, the epigenetic programming of rRNA production in somatic lineages is dependent upon continued sensing of nutrient levels via Rrn3 to generate the thrifty phenotype (Denisenko et al. 2016). Collectively, the balance between responses occurring within extra-embryonic and embryonic lineages drive the growth trajectory and later disease risk in adulthood. The maternal protein restriction model has therefore pinpointed the periconceptional period, and particularly the preimplantation embryo, as the critical stage coordinating developmental programming.

5 Maternal Obesity and Overnutrition

Whilst our understanding of how maternal undernutrition during the periconceptional period or even exclusively during the preimplantation period influences postnatal health is growing (see above), we know virtually nothing about the impact of overnutrition during this time. The chronic state of overnutrition, obesity, is on the increase worldwide and it is well-established that maternal obesity affects not only the individual's health but also their fertility and offspring health. By nature, maternal obesity and associated diverse physiological alterations will affect the oocytes before conception, the developing organism throughout pregnancy and lactation and even the gametes of the next generation which are laid down prenatally. But how sensitive is the periconceptional period to overnutrition? Considering that approximately 20% of weight-conscious women admit to binge on poor foods even if lean (this figure can rise to up to 55% in overweight women of reproductive age) (Bertoli et al. 2016), this is a relevant question which has been overlooked. With regard to maternal obesity, very few data are available with focus on the impact on the early embryo. Restricting exposure to obesity and overnutrition to around conception involves embryo manipulations before implantation (oocyte/embryo recovery, in vitro techniques, embryo transfer) which are technically demanding. Perhaps best characterised is the sensitivity of the oocyte. In oocyte donation programmes applied in human ART, recent evidence has shown that donor BMI plays a dominant role in achieving pregnancy and live birth outcomes when data are adjusted for recipient BMI and various other confounders. Chances to achieve a pregnancy and live birth progressively reduced with increasing oocyte donor BMI and were less than half if the donor was in the highest BMI quartile (average BMI 27.7) compared to the lowest BMI quartile (average BMI 19.7). Similarly, human embryos derived from obese mothers after IVF and culture display morphological and metabolic abnormalities (Leary et al. 2015; Cardozo et al. 2016; Gu et al. 2015). In bovine models, replicating conditions of an obese maternal environment during oocyte maturation in vitro resulted in similar embryonic metabolic dysfunction (Van Hoeck et al. 2011, 2013). Overall, such data indicate compromised oocyte developmental capacity of oocytes derived from obese mothers. Among mechanisms that cause these alterations, a combination of spindle abnormalities and meiotic defects, accumulation of lipids and reactive oxygen species, epigenetic aberrations and overall metabolic disturbances as a result of dysfunction of organelles such as mitochondria and ER are suggested (Dunning et al. 2014; Gu et al. 2015; Igosheva et al. 2010). In rodents, mitochondrial malfunction in the oocyte or early embryo as a consequence of maternal obesity is linked to compromised foetal and placental growth, altered mitochondrial copy number and expression of imprinted genes or nutrient transporters, even if the embryo is transferred to a non-obese foster mother (Sasson et al. 2015; McPherson et al. 2015; Luzzo et al. 2012).

The impact on offspring health of exposure to overnutrition *exclusively during the preimplantation period* has not been examined yet in vivo. Data available through approaches using in vitro culture of embryos derived from lean mothers in the presence of molecules known to be elevated systemically in maternal obesity followed by embryo transfer to lean foster mothers suggest the preimplantation period alone may be sensitive to overnutrition, similar to undernutrition (see above). For example, brief exposure of mouse morulae or blastocysts to palmitic acid in vitro results in reduced blastocyst cell number and insulin signalling. When in vitro exposed blastocysts have been transferred to control foster mothers, foetal growth restriction and postnatal catch-up growth were observed (Jungheim et al. 2011). However, we only have very limited knowledge how maternal overnutrition is translated to conditions within the reproductive tract, the environment the embryo develops in. Further open questions relate to the 'toolkit' available to the early embryo: *which sensors trigger sequences of events, are these reversible or adaptable?* An important goal, therefore, for maternal overnutrition research will be to reveal the molecular participants of communication between mother and embryo and the range of responses that the embryo can adopt and their consequences for health.

6 Conclusion

The preimplantation embryo has a full programme of developmental 'objectives' to achieve over just a few days, to establish both an intrinsic plan of multi-lineage morphogenesis and a capacity to interrogate extrinsic cues including maternal nutritional status to set the growth trajectory for pregnancy. This important responsibility can impose an imprint on offspring phenotype and health that can last a lifetime. A better understanding of both maternal and embryonic mechanisms involved in setting up the developmental trajectory early on in response to maternal nutrition is critical to develop effective intervention strategies. Perhaps even more pressing is to raise public awareness of the potential consequences of maternal malnutrition for the next generation at a time when the mother is still unaware of her pregnancy. It is this same early period in development that is the subject of numerous reproductive technologies that can also cause adverse long-term effects. Clearly, this research area should remain a priority in biomedicine to protect public health across the generations.

Acknowledgements This work was supported through awards from the Biotechnology and Biological Sciences Research Council (BB/I001840/1; BB/F007450/1), The Medical Research Council (G9800781), the NICHD National Cooperative Program (U01 HD044635) and the EU-FP7 EpiHealth and EpiHealthNet programmes to TPF, The Gerald Kerkut Charitable Trust, and NIH awards DK098817 and DK094934 to OD.

References

Artus J, Piliszek A, Hadjantonakis AK (2011) The primitive endoderm lineage of the mouse blastocyst: sequential transcription factor activation and regulation of differentiation by Sox17. Dev Biol 350(2):393–404. doi:10.1016/j.ydbio.2010.12.007

Avruch J, Long X, Ortiz-Vega S, Rapley J, Papageorgiou A, Dai N (2009) Amino acid regulation of TOR complex 1. Am J Physiol Endocrinol Metab 296(4):E592–E602. doi:10.1152/ajpendo.90645.2008

Barker D, Barker M, Fleming T, Lampl M (2013) Developmental biology: support mothers to secure future public health. Nature 504(7479):209–211

Barker DJ (2007) The origins of the developmental origins theory. J Intern Med 261(5):412–417. doi:10.1111/j.1365-2796.2007.01809.x

Barker DJ, Thornburg KL (2013) The obstetric origins of health for a lifetime. Clin Obstet Gynecol 56(3):511–519. doi:10.1097/GRF.0b013e31829cb9ca

Bartova E, Krejci J, Harnicarova A, Galiova G, Kozubek S (2008) Histone modifications and nuclear architecture: a review. J Histochem Cytochem Off J Histochem Soc 56(8):711–721. doi:10.1369/jhc.2008.951251

Beckman DA, Lloyd JB, Brent RL (1997) Investigations into mechanisms of amino acid supply to the rat embryo using whole-embryo culture. Int J Dev Biol 41(2):315–318

Bedzhov I, Graham SJ, Leung CY, Zernicka-Goetz M (2014) Developmental plasticity, cell fate specification and morphogenesis in the early mouse embryo. Philos Trans R Soc Lond B Biol Sci 369(1657):20130538. doi:10.1098/rstb.2013.0538

Bertoli S, Leone A, Ponissi V, Bedogni G, Beggio V, Strepparava MG, Battezzati A (2016) Prevalence of and risk factors for binge eating behaviour in 6930 adults starting a weight loss or maintenance programme. Public Health Nutr 19(1):71–77. doi:10.1017/S1368980015001068

Brison DR, Roberts SA, Kimber SJ (2013) How should we assess the safety of IVF technologies? Reprod Biomed Online 27(6):710–721. doi:10.1016/j.rbmo.2013.09.006

Cai KQ, Caslini C, Capo-chichi CD, Slater C, Smith ER, Wu H, Klein-Szanto AJ, Godwin AK, XX X (2009) Loss of GATA4 and GATA6 expression specifies ovarian cancer histological subtypes and precedes neoplastic transformation of ovarian surface epithelia. PLoS One 4(7):e6454. doi:10.1371/journal.pone.0006454

Capo-Chichi CD, Smedberg JL, Rula M, Nicolas E, Yeung AT, Adamo RF, Frolov A, Godwin AK, Xu XX (2010) Alteration of differentiation potentials by modulating GATA transcription factors in murine embryonic stem cells. Stem Cells Int 2010:602068. doi:10.4061/2010/602068

Cardozo ER, Karmon AE, Gold J, Petrozza JC, Styer AK (2016) Reproductive outcomes in oocyte donation cycles are associated with donor BMI. Hum Reprod 31(2):385–392. doi:10.1093/humrep/dev298

Caslini C, Capo-chichi CD, Roland IH, Nicolas E, Yeung AT, XX X (2006) Histone modifications silence the GATA transcription factor genes in ovarian cancer. Oncogene 25(39):5446–5461. doi:10.1038/sj.onc.1209533

Ceelen M, van Weissenbruch MM, Vermeiden JP, van Leeuwen FE, Delemarre-van de Waal HA (2008) Cardiometabolic differences in children born after in vitro fertilization: follow-up study. J Clin Endocrinol Metab 93(5):1682–1688. doi:10.1210/jc.2007-2432

Cha J, Sun X, Dey SK (2012) Mechanisms of implantation: strategies for successful pregnancy. Nat Med 18(12):1754–1767. doi:10.1038/nm.3012

Chen L, Wang D, Wu Z, Ma L, Daley GQ (2010) Molecular basis of the first cell fate determination in mouse embryogenesis. Cell Res 20(9):982–993. doi:10.1038/cr.2010.106

Coan PM, Vaughan OR, McCarthy J, Mactier C, Burton GJ, Constancia M, Fowden AL (2011) Dietary composition programmes placental phenotype in mice. J Physiol 589(Pt 14):3659–3670. doi:10.1113/jphysiol.2011.208629

Cockburn K, Rossant J (2010) Making the blastocyst: lessons from the mouse. J Clin Invest 120(4):995–1003. doi:10.1172/JCI41229

de Rooij SR, Roseboom TJ (2013) The developmental origins of ageing: study protocol for the Dutch famine birth cohort study on ageing. BMJ Open 3(6):e003167. doi:10.1136/bmjopen-2013-003167

Denisenko O, Lin B, Louey S, Thornburg K, Bomsztyk K, Bagby S (2011) Maternal malnutrition and placental insufficiency induce global downregulation of gene expression in fetal kidneys. J Dev Orig Health Dis 2(2):124–133. doi:10.1017/S2040174410000632

Denisenko O, Lucas ES, Sun C, Watkins AJ, Mar D, Bomsztyk K, Fleming TP (2016) Regulation of ribosomal RNA expression across the lifespan is fine-tuned by maternal diet before implantation. Biochim Biophys Acta 1859(7):906–913. doi:10.1016/j.bbagrm.2016.04.001

Dibble CC, Cantley LC (2015) Regulation of mTORC1 by PI3K signaling. Trends Cell Biol 25(9):545–555. doi:10.1016/j.tcb.2015.06.002

Diskin MG, Kenny DA (2016) Managing the reproductive performance of beef cows. Theriogenology 86(1):379–387. doi:10.1016/j.theriogenology.2016.04.052

Dodd JM, O'Brien CM, Grivell RM (2015) Modifying diet and physical activity to support pregnant women who are overweight or obese. Curr Opin Clin Nutr Metab Care 18(3):318–323. doi:10.1097/MCO.0000000000000170

Dominguez-Salas P, Moore SE, Baker MS, Bergen AW, Cox SE, Dyer RA, Fulford AJ, Guan Y, Laritsky E, Silver MJ, Swan GE, Zeisel SH, Innis SM, Waterland RA, Prentice AM, Hennig BJ (2014) Maternal nutrition at conception modulates DNA methylation of human metastable epialleles. Nat Commun 5:3746. doi:10.1038/ncomms4746

Dunning KR, Russell DL, Robker RL (2014) Lipids and oocyte developmental competence: the role of fatty acids and beta-oxidation. Reproduction 148(1):R15–R27. doi:10.1530/REP-13-0251

Eckert JJ, Fleming TP (2008) Tight junction biogenesis during early development. Biochim Biophys Acta 1778(3):717–728. doi:10.1016/j.bbamem.2007.09.031

Eckert JJ, Porter R, Watkins AJ, Burt E, Brooks S, Leese HJ, Humpherson PG, Cameron IT, Fleming TP (2012) Metabolic induction and early responses of mouse blastocyst developmental programming following maternal low protein diet affecting life-long health. PLoS One 7(12):e52791. doi:10.1371/journal.pone.0052791

Eckert JJ, Velazquez MA, Fleming TP (2015) Cell signalling during blastocyst morphogenesis. Adv Exp Med Biol 843:1–21. doi:10.1007/978-1-4939-2480-6_1

Fleming TP, Velazquez MA, Eckert JJ, Lucas ES, Watkins AJ (2012) Nutrition of females during the peri-conceptional period and effects on foetal programming and health of offspring. Anim Reprod Sci 130(3–4):193–197. doi:10.1016/j.anireprosci.2012.01.015

Fleming TP, Watkins AJ, Sun C, Velazquez MA, Smyth NR, Eckert JJ (2015) Do little embryos make big decisions? How maternal dietary protein restriction can permanently change an embryo's potential, affecting adult health. Reprod Fertil Dev 27(4):684–692. doi:10.1071/RD14455

Gardner DK, Harvey AJ (2015) Blastocyst metabolism. Reprod Fertil Dev 27:638. doi:10.1071/RD14421

Gardner DS, Pearce S, Dandrea J, Walker R, Ramsay MM, Stephenson T, Symonds ME (2004) Peri-implantation undernutrition programs blunted angiotensin II evoked baroreflex responses in young adult sheep. Hypertension 43(6):1290–1296. doi:10.1161/01.HYP.0000126991.67203.7b

Gluckman PD, Cutfield W, Hofman P, Hanson MA (2005) The fetal, neonatal, and infant environments-the long-term consequences for disease risk. Early Hum Dev 81(1):51–59. doi:10.1016/j.earlhumdev.2004.10.003

Gu L, Liu H, Gu X, Boots C, Moley KH, Wang Q (2015) Metabolic control of oocyte development: linking maternal nutrition and reproductive outcomes. Cell Mol Life Sci CMLS 72(2):251–271. doi:10.1007/s00018-014-1739-4

Hales CN, Barker DJ (2001) The thrifty phenotype hypothesis. Br Med Bull 60:5–20

Hart R, Norman RJ (2013) The longer-term health outcomes for children born as a result of IVF treatment: part I--General health outcomes. Hum Reprod Update 19(3):232–243. doi:10.1093/humupd/dms062

Heijmans BT, Tobi EW, Stein AD, Putter H, Blauw GJ, Susser ES, Slagboom PE, Lumey LH (2008) Persistent epigenetic differences associated with prenatal exposure to famine in humans. Proc Natl Acad Sci U S A 105(44):17046–17049. doi:10.1073/pnas.0806560105

Hernandez CE, Harding JE, Oliver MH, Bloomfield FH, Held SD, Matthews LR (2009) Effects of litter size, sex and periconceptional ewe nutrition on side preference and cognitive flexibility in the offspring. Behav Brain Res 204(1):82–87. doi:10.1016/j.bbr.2009.05.019

Igosheva N, Abramov AY, Poston L, Eckert JJ, Fleming TP, Duchen MR, McConnell J (2010) Maternal diet-induced obesity alters mitochondrial activity and redox status in mouse oocytes and zygotes. PLoS One 5(4):e10074. doi:10.1371/journal.pone.0010074

Jungheim ES, Louden ED, Chi MM, Frolova AI, Riley JK, Moley KH (2011) Preimplantation exposure of mouse embryos to palmitic acid results in fetal growth restriction followed by catch-up growth in the offspring. Biol Reprod 85(4):678–683. doi:10.1095/biolreprod.111.092148

Kermack AJ, Finn-Sell S, Cheong YC, Brook N, Eckert JJ, Macklon NS, Houghton FD (2015) Amino acid composition of human uterine fluid: association with age, lifestyle and gynaecological pathology. Hum Reprod 30(4):917–924. doi:10.1093/humrep/dev008

Kim E (2009) Mechanisms of amino acid sensing in mTOR signaling pathway. Nutr Res Pract 3(1):64–71. doi:10.4162/nrp.2009.3.1.64

Kwong WY, Wild AE, Roberts P, Willis AC, Fleming TP (2000) Maternal undernutrition during the preimplantation period of rat development causes blastocyst abnormalities and programming of postnatal hypertension. Development 127(19):4195–4202

Lazaraviciute G, Kauser M, Bhattacharya S, Haggarty P (2014) A systematic review and meta-analysis of DNA methylation levels and imprinting disorders in children conceived by IVF/ICSI compared with children conceived spontaneously. Hum Reprod Update 20(6):840–852. doi:10.1093/humupd/dmu033

Leary C, Leese HJ, Sturmey RG (2015) Human embryos from overweight and obese women display phenotypic and metabolic abnormalities. Hum Reprod 30(1):122–132. doi:10.1093/humrep/deu276

Leese HJ (2012) Metabolism of the preimplantation embryo: 40 years on. Reproduction 143(4):417–427. doi:10.1530/REP-11-0484

Leroy JL, Valckx SD, Jordaens L, De Bie J, Desmet KL, Van Hoeck V, Britt JH, Marei WF, Bols PE (2015) Nutrition and maternal metabolic health in relation to oocyte and embryo quality: critical views on what we learned from the dairy cow model. Reprod Fertil Dev 27(4):693–703. doi:10.1071/RD14363

Li L, Lu X, Dean J (2013) The maternal to zygotic transition in mammals. Mol Aspects Med 34(5):919–938. doi:10.1016/j.mam.2013.01.003

Liu X, Zhao D, Zheng Y, Wang L, Qian Y, Xu C, Huang H, Hwa YL, Jin F (2014) Expression of histone acetyltransferase GCN5 and histone deacetylase 1 in the cultured mouse preimplantation embryos. Curr Pharm Des 20(11):1772–1777

Luzzo KM, Wang Q, Purcell SH, Chi M, Jimenez PT, Grindler N, Schedl T, Moley KH (2012) High fat diet induced developmental defects in the mouse: oocyte meiotic aneuploidy and fetal growth retardation/brain defects. PLoS One 7(11):e49217. doi:10.1371/journal.pone.0049217

McEvoy TG, Sinclair KD, Young LE, Wilmut I, Robinson JJ (2000) Large offspring syndrome and other consequences of ruminant embryo culture in vitro: relevance to blastocyst culture in human ART. Hum Fertil (Camb) 3(4):238–246

McPherson NO, Bell VG, Zander-Fox DL, Fullston T, LL W, Robker RL, Lane M (2015) When two obese parents are worse than one! Impacts on embryo and fetal development. Am J Physiol Endocrinol Metab 309(6):E568–E581. doi:10.1152/ajpendo.00230.2015

Moore SE, Cole TJ, Poskitt EM, Sonko BJ, Whitehead RG, McGregor IA, Prentice AM (1997) Season of birth predicts mortality in rural Gambia. Nature 388(6641):434. doi:10.1038/41245

Morrisey EE, Musco S, Chen MY, MM L, Leiden JM, Parmacek MS (2000) The gene encoding the mitogen-responsive phosphoprotein Dab2 is differentially regulated by GATA-6 and GATA-4 in the visceral endoderm. J Biol Chem 275(26):19949–19954. doi:10.1074/jbc.M001331200

Nardelli AA, Stafinski T, Motan T, Klein K, Menon D (2014) Assisted reproductive technologies (ARTs): evaluation of evidence to support public policy development. Reprod Health 11(1):76. doi:10.1186/1742-4755-11-76

Painter RC, de Rooij SR, Bossuyt PM, Simmers TA, Osmond C, Barker DJ, Bleker OP, Roseboom TJ (2006) Early onset of coronary artery disease after prenatal exposure to the Dutch famine. Am J Clin Nutr 84(2):322–327. quiz 466-327

Rivera RM, Ross JW (2013) Epigenetics in fertilization and preimplantation embryo development. Prog Biophys Mol Biol 113(3):423–432. doi:10.1016/j.pbiomolbio.2013.02.001

Roseboom TJ, Painter RC, van Abeelen AF, Veenendaal MV, de Rooij SR (2011) Hungry in the womb: what are the consequences? Lessons from the Dutch famine. Maturitas 70(2):141–145. doi:10.1016/j.maturitas.2011.06.017

Rossant J, Chazaud C, Yamanaka Y (2003) Lineage allocation and asymmetries in the early mouse embryo. Philos Trans R Soc Lond B Biol Sci 358(1436):1341–1348.; discussion 1349. doi:10.1098/rstb.2003.1329

Sasson IE, Vitins AP, Mainigi MA, Moley KH, Simmons RA (2015) Pre-gestational vs gestational exposure to maternal obesity differentially programs the offspring in mice. Diabetologia 58(3):615–624. doi:10.1007/s00125-014-3466-7

Schrode N, Saiz N, Di Talia S, Hadjantonakis AK (2014) GATA6 levels modulate primitive endoderm cell fate choice and timing in the mouse blastocyst. Dev Cell 29(4):454–467. doi:10.1016/j.devcel.2014.04.011

Seshagiri PB, Sen Roy S, Sireesha G, Rao RP (2009) Cellular and molecular regulation of mammalian blastocyst hatching. J Reprod Immunol 83(1–2):79–84. doi:10.1016/j.jri.2009.06.264

Sim KA, Partridge SR, Sainsbury A (2014) Does weight loss in overweight or obese women improve fertility treatment outcomes? A systematic review. Obes Rev Off J Int Assoc Stud Obes 15(10):839–850. doi:10.1111/obr.12217

Sozen B, Can A, Demir N (2014) Cell fate regulation during preimplantation development: a view of adhesion-linked molecular interactions. Dev Biol 395(1):73–83. doi:10.1016/j.ydbio.2014.08.028

Sun C, Denisenko O, Sheth B, Cox A, Lucas ES, Smyth NR, Fleming TP (2015) Epigenetic regulation of histone modifications and Gata6 gene expression induced by maternal diet in mouse embryoid bodies in a model of developmental programming. BMC Dev Biol 15(1):3. doi:10.1186/s12861-015-0053-1

Sun C, Velazquez MA, Marfy-Smith S, Sheth B, Cox A, Johnston DA, Smyth N, Fleming TP (2014) Mouse early extra-embryonic lineages activate compensatory endocytosis in response to poor maternal nutrition. Development 141(5):1140–1150. doi:10.1242/dev.103952

Todd SEOM, Jaquiery AL, Bloomfield FH, Harding JE (2009) Periconceptional undernutrition of ewes impairs glucose tolerance in their adult offspring. Pediatr Res 65:409–413

Torrens C, Snelling TH, Chau R, Shanmuganathan M, Cleal JK, Poore KR, Noakes DE, Poston L, Hanson MA, Green LR (2009) Effects of pre- and periconceptional undernutrition on arterial function in adult female sheep are vascular bed dependent. Exp Physiol 94(9):1024–1033. doi:10.1113/expphysiol.2009.047340

Turner N, Robker RL (2015) Developmental programming of obesity and insulin resistance: does mitochondrial dysfunction in oocytes play a role? Mol Hum Reprod 21(1):23–30. doi:10.1093/molehr/gau042

Van Hoeck V, Leroy JL, Arias Alvarez M, Rizos D, Gutierrez-Adan A, Schnorbusch K, Bols PE, Leese HJ, Sturmey RG (2013) Oocyte developmental failure in response to elevated nonesterified fatty acid concentrations: mechanistic insights. Reproduction 145(1):33–44. doi:10.1530/REP-12-0174

Van Hoeck V, Sturmey RG, Bermejo-Alvarez P, Rizos D, Gutierrez-Adan A, Leese HJ, Bols PE, Leroy JL (2011) Elevated non-esterified fatty acid concentrations during bovine oocyte maturation compromise early embryo physiology. PLoS One 6(8):e23183. doi:10.1371/journal.pone.0023183

Waterland RA, Kellermayer R, Laritsky E, Rayco-Solon P, Harris RA, Travisano M, Zhang W, Torskaya MS, Zhang J, Shen L, Manary MJ, Prentice AM (2010) Season of conception in rural gambia affects DNA methylation at putative human metastable epialleles. PLoS Genet 6(12):e1001252. doi:10.1371/journal.pgen.1001252

Watkins AJ, Lucas ES, Marfy-Smith S, Bates N, Kimber SJ, Fleming TP (2015) Maternal nutrition modifies trophoblast giant cell phenotype and fetal growth in mice. Reproduction 149(6):563–575. doi:10.1530/REP-14-0667

Watkins AJ, Lucas ES, Torrens C, Cleal JK, Green L, Osmond C, Eckert JJ, Gray WP, Hanson MA, Fleming TP (2010) Maternal low-protein diet during mouse pre-implantation development induces vascular dysfunction and altered renin-angiotensin-system homeostasis in the offspring. Br J Nutr 103(12):1762–1770. doi:10.1017/S0007114509993783

Watkins AJ, Lucas ES, Wilkins A, Cagampang FR, Fleming TP (2011) Maternal periconceptional and gestational low protein diet affects mouse offspring growth, cardiovascular and adipose phenotype at 1 year of age. PLoS One 6(12):e28745. doi:10.1371/journal.pone.0028745

Watkins AJ, Sinclair KD (2014) Paternal low protein diet affects adult offspring cardiovascular and metabolic function in mice. Am J Physiol Heart Circ Physiol 306(10):H1444–H1452. doi:10.1152/ajpheart.00981.2013

Watkins AJ, Ursell E, Panton R, Papenbrock T, Hollis L, Cunningham C, Wilkins A, Perry VH, Sheth B, Kwong WY, Eckert JJ, Wild AE, Hanson MA, Osmond C, Fleming TP (2008a) Adaptive responses by mouse early embryos to maternal diet protect fetal growth but predispose to adult onset disease. Biol Reprod 78(2):299–306. doi:10.1095/biolreprod.107.064220

Watkins AJ, Wilkins A, Cunningham C, Perry VH, Seet MJ, Osmond C, Eckert JJ, Torrens C, Cagampang FR, Cleal J, Gray WP, Hanson MA, Fleming TP (2008b) Low protein diet fed exclusively during mouse oocyte maturation leads to behavioural and cardiovascular abnormalities in offspring. J Physiol 586(8):2231–2244. doi:10.1113/jphysiol.2007.149229

Zhou LQ, Dean J (2015) Reprogramming the genome to totipotency in mouse embryos. Trends Cell Biol 25(2):82–91. doi:10.1016/j.tcb.2014.09.006

Zohn IE, Sarkar AA (2010) The visceral yolk sac endoderm provides for absorption of nutrients to the embryo during neurulation. Birth Defects Res A Clin Mol Teratol 88(8):593–600. doi:10.1002/bdra.20705

The Long-Term Effect of the Periconception Period on the Embryo's Epigenetic Profile and Phenotype: The Role of Maternal Disease Such as Diabetes and How the Effect Is Mediated (Example from a Rabbit Model)

Bernd Fischer, Maria Schindler, S. Mareike Pendzialek, Jacqueline Gürke, Elisa Haucke, Katarzyna Joanna Grybel, René Thieme, and Anne Navarrete Santos

Abstract Maternal metabolic diseases such as diabetes mellitus with diabetogenic hypoinsulinemia and hyperglycemia change periconceptional developmental conditions in utero. In preimplantation rabbit embryos, all major metabolic pathways are affected. Alterations in protein, lipid and glucose metabolism, adipokines, advanced glycation end products (AGEs) and reactive oxygen species (ROS) are described in this review. The embryonic metabolism is characterized by a high plasticity which enables survival of most preimplantation embryos under the non-physiological developmental conditions in diabetic mothers. Adiponectin, for example, compensates for the missing insulin-driven glucose supply and stimulates intracellular lipid accumulation in embryonic cells. AGEs and ROS are clear indicators of metabolic stress. The price paid for survival, however, needs to be taken into consideration. It is an increase in lipogenesis and proteinogenesis, leading to metabolic stress and with potentially negative long-term health effects.

Keywords Diabetes • Embryo metabolism • Adiponectin • Advanced glycation end products (AGE) • Reactive oxygen species (ROS)

B. Fischer (✉) • M. Schindler • S. Mareike Pendzialek • J. Gürke • E. Haucke • K.J. Grybel • R. Thieme • A.N. Santos
Department of Anatomy and Cell Biology, Martin Luther University Faculty of Medicine, Grosse Steinstrasse 52, D - 06114 Halle, Germany
e-mail: bernd.fischer@medizin.uni-halle.de

A. Fazeli, W.V. Holt (eds.), *Periconception in Physiology and Medicine*, Advances in Experimental Medicine and Biology 1014,
DOI 10.1007/978-3-319-62414-3_6

1 Introduction

In Germany more than 35.350 newborns are affected by maternal diabetes mellitus each year, as approximately 6% of all pregnancies occur in diabetic women (Deutscher Gesundheitsbericht Diabetes 2015). Amongst these women, 18% suffer from a preconceptional diabetes. The majority is affected by gestational diabetes. In the USA, 15% of pregnancies in women are diabetic (International Diabetes Foundation Atlas 2015). On an international scale, similar numbers are reported with approximately 17% of all births being affected by hyperglycemia during pregnancy (International Diabetes Foundation Atlas 2015). Again, gestational diabetes stands in most frequently (approximately 85%) followed by pre-existing type I diabetes (7%) and type II diabetes (5%) (Aceti et al. 2012). In the UK, between 1996 and 2004, the number of pregnancies complicated by pre-existing diabetes increased by 50% and the prevalence of gestational diabetes mellitus doubled (Aceti et al. 2012).

Diabetes during pregnancy is a significant medical challenge. It is associated with pregnancy complications, perinatal pathologies and a higher risk for the offspring to suffer from metabolic diseases later in life. Diabetic women are sub-fertile as indicated by ovulation disorders and higher risks for early spontaneous miscarriages (Casson et al. 1997, Penney et al. 2003, Verheijen et al. 2005, Yang et al. 2006). On average, the rate of pre-term births amongst women with pre-existing diabetes is about four times higher compared to women without diabetes. It is well known to obstetricians and pediatricians that about half of the newborns from diabetic mothers suffer from metabolic and adaptational dysfunctions after birth. The less controlled diabetes is, the more babies are born with congenital anomalies in women with pre-existing diabetes. Children born to diabetic mothers have higher risks of congenital malformations and childhood obesity. They have a predisposition to develop type II diabetes and cardiovascular diseases (metabolic syndrome) later in life. The National Institute for Health and Clinical Excellence in the UK advised diabetic women whose glycated haemoglobin A (HbA1c) is above 10% to avoid pregnancy. It is noteworthy that diabetes type II is no longer the disease of the elderly (Diabetes Atlas Update, International Diabetes Federation 2015). An increasing number of younger women who have not given birth so far develop diabetes. Therefore the number of diabetic pregnancies will rise in the future.

These data clearly demonstrate that a diabetic pregnancy requires special medical care and attention—for both the mother and her newborn. The health consequences for children from diabetic mothers raise the questions about when the metabolic and developmental disorders start. That the periconceptional (and preimplantation) period is a specifically critical period of pregnancy has convincingly been shown by the group of T.P. Fleming from Southampton. They have fed mice a protein-reduced diet during the short periconceptional time and describe numerous health effects in offspring later in life (Watkins et al. 2008a, b, 2010, 2011). The current review focuses on diabetogenic perturbations of preimplantation embryo development in a rabbit diabetic model. We asked which potential mechanisms are involved in diabetogenic disorders and when they become noticeable in embryonic development.

2 The Diabetic Rabbit Model

The study of diabetogenic disorders in embryos cannot be performed in women. Therefore, valid experimental animal models are needed. Basically, diabetes mellitus can be experimentally induced by destruction of the pancreatic beta cells by chemicals (such as alloxan) or in genetic models, i.e. in animals with a naturally occurring diabetes, or by genetic modification, for example, in pigs as recently reported (Wolf et al. 2014). The described rabbit model is a diabetes type 1 model as maternal endogenous insulin production rapidly ceases after alloxan treatment. The review will summarize data raised in the rabbit model with a specific focus on the pre- and peri-implantation period.

The diabetic model is shown in Fig. 1. Sexually mature female rabbits are treated with alloxan. This chemical specifically destroys the pancreatic B cells leading to an experimentally induced type I-like diabetes (Ramin et al. 2010, Thieme et al. 2012a, b, Schindler et al. 2013, 2014, Gürke et al. 2015, 2016). The rabbit is used as an experimental model as its embryology resembles that of the human in many aspects (Fischer et al. 2012). The rabbit is a reflex ovulating species allowing easy scheduling of experiments involving stage- and age-specific maternal tissues and embryo analyses, even as close as in hours post coitum (pc). Stimulation of folliculogenesis (superovulation) leads to, on average, blastocyst numbers above 20. However, the number of corpora lutea and embryo recovery rates at day 6 pc are lower in diabetic than in healthy donor rabbits (Ramin et al. 2010), reflecting the lower fertility of diabetic females compared with healthy ones.

Blastocysts can be flushed out of the uteri up to the primitive streak stage (stage III), as implantation starts late on day 6 pc at stage IV. At day 6 pc, the rabbit blastocyst is 2.5–3 mm in size, allowing retrieval of substantial material for analyses and isolation of the embryonic cell lineages by microdissection. In a day 6 pc blastocyst, approximately 2,000–7,000 embryoblast cells, depending on the stage of development, and at least ten times more trophoblast cells (based on protein measurements), can be separated and recovered for further processing. This is a unique advantage of the rabbit model, saving on one side animal experiments and donor rabbits and allowing, on the other, subtle analyses in individual blastocysts.

After establishment of diabetes for 6–10 days, follicle stimulation and mating, preimplantation embryos are recovered, usually in the blastocyst stage at day 6 pc. Maternal blood glucose levels are kept between 19 and 30 mM by insulin supplementation thrice daily. Each time, insulin supplementation is adjusted according to the measured blood glucose level.

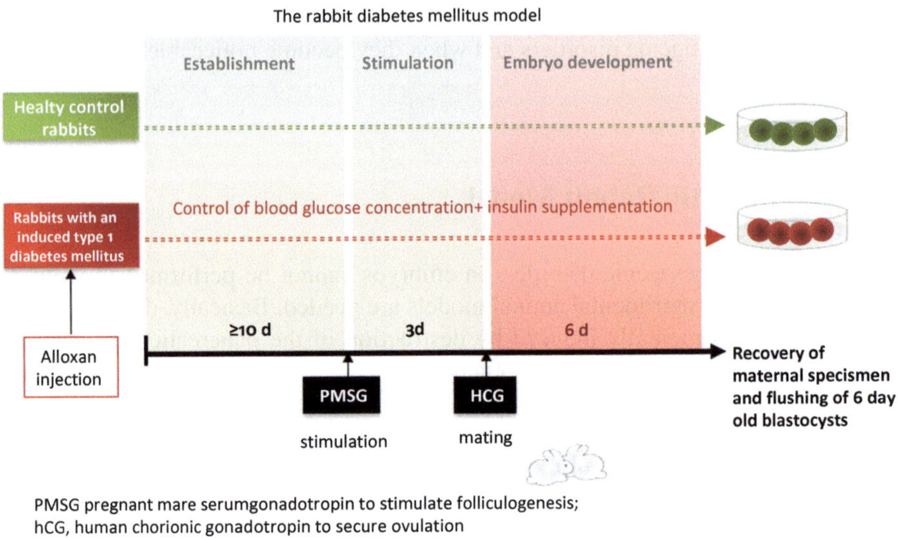

Fig. 1 The diabetic rabbit model. Experimental schedule. Blood glucose levels in alloxan-treated rabbits were kept between 25 and 30 mmol/l by insulin supplementation twice daily. Blood glucose in healthy rabbits is about 5 mmol/l

3 Metabolic Changes in Diabetic Rabbits and Their Blastocysts

3.1 Insulin and Glucose

First, we studied the insulin and glucose concentrations in the alloxan-treated females and their offspring. In diabetic rabbits, the plasma insulin concentrations were decreased by 85% to 15 pM. As rabbit blastocysts do not express insulin and due to the dramatic decrease in maternal insulin, no insulin could be identified in day 6 pc blastocysts recovered from diabetic mothers (Ramin et al. 2010). The measured uterine glucose concentrations reflected the plasma glucose concentrations in healthy and diabetic rabbits, respectively, albeit on a different scale. In diabetic rabbits, the plasma glucose concentration was kept to about 25 mM (see Fig. 1; glucose levels in controls approximately 5 mM). In the uterine cavity of such females, the glucose levels were clearly lower than in plasma but about three times higher than in healthy controls (1.75 mM compared to 0.5 mM in controls; Ramin et al. 2010). Therefore, metabolic disorders like missing insulin and hyperglycemia do reach the embryo and its developmental milieu right from the start at conception.

In plasma and uterine fluid and in blastocysts from diabetic rabbits, various significant metabolic changes are noticeable. As a more general observation, two major findings should be mentioned first. Blastocysts grown in a diabetic mother are retarded in development (Ramin et al. 2010, Thieme et al. 2012a, b). Secondly, diabetes changes major metabolic pathways in the mother and in the embryo. So far, all pathways being investigated were altered. These changes, however, were associated with various compensatory actions by the mother and the embryo. For example, the preimplantation embryo compensates the loss of insulin and systemic IGF I by an enhanced endogenous IGF I and IGF II expression (Thieme et al. 2012 b). It needs to be emphasized that most embryos survive and continue development. However, further development is based on an altered embryonic metabolism compared to normal development.

3.2 Adipokines

Adipokines (such as adiponectin; Schindler et al. 2013) and triglycerides (Schindler et al. 2014) are clearly increased in the blood of diabetic rabbits. In their blastocysts, adiponectin and the adiponectin receptor 1 (AdipoR1) expression is enhanced. If the blastocysts' cell lineages are compared, adiponectin expression is particularly stronger in trophoblast cells of diabetic blastocysts. Adiponectin stimulates blastocyst glucose uptake in vitro and may substitute insulin action in glucose uptake, for example, by GLUT4 translocation into the apical cell membrane in trophoblast cells (Fischer et al. 2010).

3.3 Lipid Metabolism

An astonishing and striking observation in our studies was a strongly increased accumulation of intracellular lipid droplets and lipogenic marker genes in embryoblast and trophoblast cells from diabetic blastocysts (Schindler et al. 2014). The increase in adiponectin expression could be involved in this effect. Adiponectin may play a crucial role in glucose uptake and altered lipid metabolism in diabetic rabbit blastocysts. Looking at lipogenic genes involved, for example, FABP4, FATP4, CPT1, SREBP1 and PPARs (Schindler et al. 2014, Schindler et al. unpublished), it becomes obvious that major lipid metabolism pathways such as fatty acid uptake and beta-oxidation are affected. Adiponectin may serve as a failsafe regulation for blastocysts surviving in a diabetic uterine milieu. However, survival is not the same as in normal development, as shown by metabolic changes and intracellular accumulation of lipids due to compensatory actions required for embryo survival. We consider these observations as good experimental proof of evidence for the developmental origin of health and diseases (DOHaD) theorem. The rabbit model has provided new and constant results. All metabolic pathways studied so far in rabbit embryos were changed. Importantly, these changes ("origins") start at a very early stage of development.

3.4 Amino Acids and AGEs

The branched chain amino acids (BCAA) leucine, isoleucine and valine were increased by a factor of two in maternal plasma, uterine fluid and in blastocysts compared to healthy controls (Gürke et al. 2015). The higher availability of BCAA induced embryonic mTORC1 (mechanistic/mammalian target of rapamycin complex 1) signalling (Gürke et al. 2015, 2016). mTOR is a cellular nutrient sensor and master regulator of protein metabolism. Enhanced mTORC1 signalling increased protein synthesis in day 6 rabbit blastocysts, particularly in trophoblast cells (Gürke et al. 2016), leading to another important consequence. These findings demonstrate that non-physiological developmental conditions affect embryo development in a cell lineage-specific manner. As the ontogenetic impact of embryoblast (ICM) and trophoblast varies stage-specifically during further development, a precise description of these changes is needed to better understand diabetogenic disorders in pre- and postnatal development.

An increase in advanced glycation end products is a well-known phenomenon in diabetics. AGEs are formed non-enzymatically by reactions between reducing sugars and amine groups on proteins. A diabetic environment leads to an increased protein oxidation and AGE formation (Buongiorno et al. 1997, Guosheng et al. 2009), as confirmed in our model. In rabbits, AGEs were increased in blood plasma of diabetic females (up to 50%), correlating closely with an AGE accumulation in the endometrium and in blastocyst cavity fluid. These increases were accompanied by an increased RAGE (receptor of AGE) mRNA expression in blastocysts (Haucke et al. 2014). The increased AGEs constitute a remarkable metabolic stress for embryonic cells. A proposed mechanism of AGE-induced cellular damage is the release of reactive oxygen species, particularly superoxide and hydrogen peroxide. ROS are naturally occurring in mammalian cells; however, an excess of ROS generation overloads the cells' anti-oxidant defence mechanisms and in turn leads to oxidative stress. This, again, was detected through the use of the indicator 2′, 7′-dichlorofluorescein-diacetate (DCF-DA). Employing the diabetic rabbit model, we demonstrated clearly higher ROS levels in blastocysts from diabetic than from healthy females (Fig. 2). Recently, it has been shown that offspring from mothers with polycystic ovary syndrome (PCOS) and gestational diabetes showed increased oxidative stress markers (Boutzios et al. 2013).

These metabolic changes occur at a very early and vulnerable ontogenetic developmental stage and may persist. An increase in AGEs would implicate long-term effects on further prenatal and postnatal development of embryos from diabetic mothers.

To conclude, maternal metabolic diseases such as the diabetogenic hypoinsulinemia and hyperglycemia significantly change periconceptional and preimplantation developmental conditions in utero. In preimplantation rabbit embryos, all major metabolic pathways are affected on various regulatory levels (microRNA (not discussed), mRNA, protein). Preimplantation rabbit embryos adapt and compensate

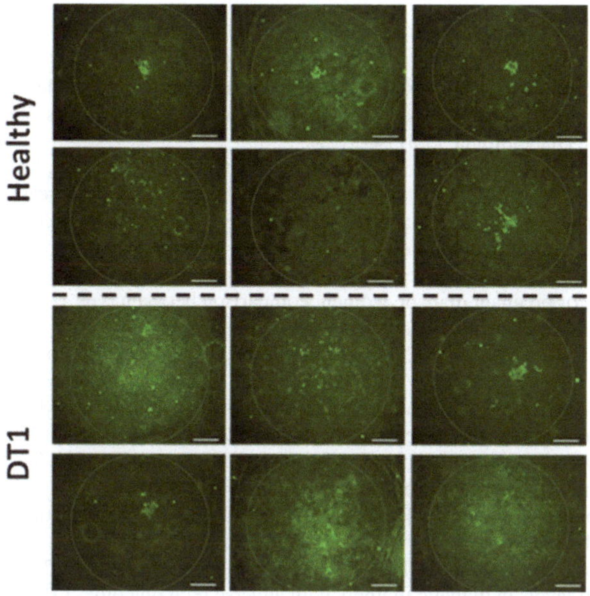

Fig. 2 Intracellular reactive oxygen species (ROS) in diabetic rabbit blastocysts. ROS measurements are shown in 6-day-old rabbit blastocysts developed under maternal diabetes mellitus type 1 (*DT1*) compared to the healthy control reference group (healthy). Blastocysts were stained with 2′, 7′-dichlorofluorescein-diacetate (DCFH-DA) and examined by fluorescence microscopy. Twelve individual blastocysts are shown ($N = 2$, $n = 6$). The area inside the circle marks embryoblast cells, outside trophoblast cells. Bars = 200 μm

for the non-physiological developmental conditions. The embryonic metabolism is characterized by a high plasticity which allows metabolic adaption and enables survival of most preimplantation embryos in a diabetic mother. Mechanisms identified are, amongst others, a hormonal switch from an insulin to an adiponectin-regulated metabolism with consequences for embryonic lipid and amino acid metabolism and accompanied by increased AGE and ROS levels.

The rabbit model shows that most mammalian embryos survive a diabetic pregnancy due to the high metabolic plasticity of embryonic cells, successfully overcoming cellular stress and damage, however, with potential long-term negative effects on pre- and postnatal development and the offspring's health.

Acknowledgements The work described was funded by EU (FP7-EpiHealth (No. 278418), EpiHealthNet (No. 317146), Cost Actions TD 1101 (RGB-Net), FA1201 (EPICONCEPT) and BM 1308 (SALAAM)), the Wilhelm Roux Programme of the MLU Faculty of Medicine and the German Research Council (DFG (NA 418/4-2)).

References

Aceti A, Santhakumaran S, Logan KM, Philipps LH, Prior E, Gale C, Hyde MJ, Modi N (2012) The diabetic pregnancy and offspring blood pressure in childhood: a systematic review and meta-analysis. Diabetologia 55:3114–3127

Boutzios G, Livadas S, Piperi C, Vitoratos N, Adamopoulos C, Hassiakos D, Iavazzo C, Diamanti-Kandarakis E (2013) Polycystic ovary syndrome offspring display increased oxidative stress markers comparable to gestational diabetes offspring. Fertil Steril 99:943–950

Buongiorno AM, Morelli S, Sagratella E, Castaldo P, Di Virgilio A, Maroccia E, Ricciardi G, Sciullo E, Cardellini G, Fallucca F (1997) Levels of advanced glycosylation end-products (AGE) in sera of pregnant diabetic women: comparison between type 1, type 2 and gestational diabetes mellitus. Annali dell'Istituto Superiore di Sanita 33:375–378

Casson IF, Clarke CA, Howard CV, McKendrick O, Pennycook S, Pharoah PO, Platt MJ, Stanisstreet M, van Velszen D, Walkinshaw S (1997) Outcomes of pregnancy in insulin dependent diabetic women: results of a five year population cohort study. Br Med J 315:275–278

Deutscher Gesundheitsbericht Diabetes, diabetesDE – Deutsche Diabetes-Hilfe (2015)

Fischer B, Chavatte-Palmer P, Viebahn C, Navarrete Santos A, Duranthon V (2012) Rabbit as a reproductive model for human health. Reproduction 144:1–10

Fischer S, Santos AN, Thieme R, Ramin N, Fischer B (2010) Adiponectin stimulates glucose uptake in rabbit blastocysts. Biol Reprod 83:859–865

Guosheng L, Hongmei S, Chuan N, Haiying L, Xiaopeng Z, Xianqiong L (2009) The relationship of serum AGE levels in diabetic mothers with adverse fetal outcome. J Perinatol 29:483–488

Gürke J, Hirche F, Thieme R, Haucke E, Schindler M, Stangl GI, Fischer B, Navarrete Santos A (2015) Maternal diabetes leads to adaptation in embryonic amino acid metabolism during early pregnancy. PLoS One 10:e0127465

Gürke J, Schindler M, Pendzialek SM, Thieme R, Grybel KJ, Heller R, Spengler K, Fleming TP, Fischer B, Navarrete Santos A (2016) Maternal diabetes promotes mTORC1 downstream signalling in rabbit preimplantation embryos. Reproduction 151:465–476

Haucke E, Navarrete Santos A, Simm A, Henning C, Glomb MA, Gürke J, Schindler M, Fischer B, Navarrete Santos A (2014) Accumulation of advanced glycation end products in the rabbit blastocyst under maternal diabetes. Reproduction 148:169–178

International Diabetes Foundation Atlas, Seventh edition, 2015

Penney GC, Mair G, Pearson DWM, Scottish Diabetes in Pregnancy Group (2003) Outcomes of pregnancies in women with type 1 diabetes in Scotland: a national population-based study. International Journal of Obstetrics and Gynaecology 110:315–318

Ramin N, Thieme R, Fischer S, Schindler M, Schmidt T, Fischer B, Navarrete Santos A (2010) Maternal diabetes impairs gastrulation and insulin and IGF-I receptor expression in rabbit blastocysts. Endocrinology 151:4158–4167

Schindler M, Fischer S, Thieme R, Fischer B, Santos AN (2013) cAMP-responsive element binding protein: a vital link in embryonic hormonal adaptation. Endocrinology 154:2208–2221

Schindler M, Pendzialek M, Navarrete Santos A, Plösch T, Seyring S, Gürke J, Haucke E, Knelangen JM, Fischer B, Santos AN (2014) Maternal diabetes leads to unphysiological high lipid accumulation in rabbit preimplantation embryos. Endocrinology 155:1498–1509

Thieme R, Ramin N, Fischer S, Püschel B, Fischer B, Santos AN (2012a) Gastrulation in rabbit blastocysts depends on insulin and insulin-like-growth-factor 1. Mol Cell Endocrinol 348:112–119

Thieme R, Schindler M, Ramin N, Fischer S, Mühleck B, Fischer B, Navarrete Santos A (2012b) Insulin growth factor adjustment in preimplantation rabbit blastocysts and uterine tissues in response to maternal type 1 diabetes. Mol Cell Endocrinol 358:96–103

Verheijen ECJ, Critchley JA, Whitelaw DC, Tuffnell DJ (2005) Outcomes of pregnancies in women with pre-existing type 1 or type 2 diabetes, in an ethnically mixed population. Int J Obstet Gynaecol 112:1500–1503

Watkins AJ, Lucas ES, Torrens C, Cleal JK, Green L, Osmond C, Eckert JJ, Gray WP, Hanson MA, Fleming TP (2010) Maternal protein restriction during mouse preimplantation development induces offspring vascular dysfunction and alters renin-angiotensin-system homeostasis. Br J Nutr 103:1762–1770

Watkins AJ, Lucas ES, Wilkins A, Cagampang FR, Fleming TP (2011) Maternal periconceptional and gestational low protein diet affects mouse offspring growth, cardiovascular and adipose phenotype at 1 year of age. PLoS One 6(12):e28745

Watkins AJ, Ursell E, Panton R, Papenbrock T, Hollis L, Cunningham C, Wilkins A, Perry VH, Sheth B, Kwong WY, Eckert JJ, Wild AE, Hanson MA, Osmond C, Fleming TP (2008a) Adaptive responses by mouse early embryos to maternal diet protect fetal growth but predispose to adult onset disease. Biol Reprod 78:299–306

Watkins AJ, Wilkins A, Cunningham C, Perry VH, Seet MJ, Osmond C, Eckert JJ, Torrens C, Cagampang FR, Cleal J, Gray WP, Hanson MA, Fleming TP (2008b) Low protein diet fed exclusively during mouse oocyte maturation leads to behavioural and cardiovascular abnormalities in offspring. J Physiol 586:2231–2244

Wolf E, Braun-Reichhart C, Streckel E, Renner S (2014) Genetically engineered pig models for diabetes research. Transgenic Res 23:27–38

Yang J, Cummings EA, O'connell C, Jangaard K (2006) Fetal and neonatal outcomes of diabetic pregnancies. Obstet Gynecol 108:644–650

Long-Term Effects of the Periconception Period on Embryo Epigenetic Profile and Phenotype: The Role of Stress and How This Effect Is Mediated

James Ord, Alireza Fazeli, and Penelope J. Watt

Abstract Stress represents an unavoidable aspect of human life, and pathologies associated with dysregulation of stress mechanisms – particularly psychiatric disorders – represent a significant global health problem. While it has long been observed that levels of stress experienced in the periconception period may greatly affect the offspring's risk of psychiatric disorders, the mechanisms underlying these associations are not yet comprehensively understood. In order to address this question, this chapter will take a 'top-down' approach, by first defining stress and associated concepts, before exploring the mechanistic basis of the stress response in the form of the hypothalamic-pituitary-adrenal (HPA) axis, and how dysregulation of the HPA axis can impede our mental and physical health, primarily via imbalances in glucocorticoids (GCs) and their corresponding receptors (GRs) in the brain. The current extent of knowledge pertaining to the impact of stress on developmental programming and epigenetic inheritance is then extensively discussed, including the role of chromatin remodelling associated with specific HPA axis-related genes and the possible role of regulatory RNAs as messengers of environmental stress both in the

J. Ord
Department of Animal and Plant Sciences, University of Sheffield, Western Bank, Sheffield S10 2TN, UK

Academic Unit of Reproductive and Developmental Medicine, University of Sheffield, Level 4, Jessop Wing, Tree Root Walk, Sheffield S10 2SF, UK

A. Fazeli (✉)
Academic Unit of Reproductive and Developmental Medicine, University of Sheffield, Level 4, Jessop Wing, Tree Root Walk, Sheffield S10 2SF, UK

Institute of Biomedicine and Translational Medicine, Department of Pathophysiology, University of Tartu, 14B Ravila, Tartu 50411, Estonia
e-mail: A.Fazeli@sheffield.ac.uk; Alireza.Fazeli@ut.ee

P.J. Watt
Department of Animal and Plant Sciences, University of Sheffield, Western Bank, Sheffield S10 2TN, UK

A. Fazeli, W.V. Holt (eds.), *Periconception in Physiology and Medicine*, Advances in Experimental Medicine and Biology 1014,
DOI 10.1007/978-3-319-62414-3_7

intrauterine environment and across the germ line. Furthering our understanding of the role of stress on embryonic development is crucial if we are to increase our predictive power of disease risk and devise-effective treatments and intervention strategies.

Keywords Stress • Behaviour • Periconception • Hypothalamic-pituitary-adrenal (HPA) axis • Glucocorticoids • Psychiatric disorders • microRNAs

1 Introduction

Psychiatric disorders such as anxiety, depression, schizophrenia, post-traumatic stress disorder (PTSD), and autism spectrum disorder (ASD) represent an enormous source of human suffering and are one of the leading causes of disability (Kalueff et al. 2014; Vos et al. 2015). While these conditions have a broad range of effects on our cognition, awareness, mood, and our perception of reality, they all implicate dysregulation of our biological stress response system: the hypothalamic-pituitary-adrenal (HPA) axis. Responding appropriately to stressful situations is integral to our survival, but the mechanism by which we do so can be impaired, to the detriment of both our psychological and physical health.

Both genetic and environmental factors are likely to contribute to the risk of developing psychiatric disorders. The environmental contribution to risk is not only affected by our own experiences but, as a growing body of evidence now suggests, by our parents' experiences either during pregnancy or even before conception. Epidemiological data reveal that children whose mothers experience stress during pregnancy are at a higher risk of psychiatric disorders (Khashan et al. 2008), whilst rodent models show that gestational stress increases anxiety-like behaviour in adulthood (Lupien et al. 2009). The topic is further complicated by the interplay between stress and other aspects of our health; physical and psychological health often seem to go hand in hand, with patients suffering from psychological disorders at higher risk of detriment to their physical health, and vice versa (Bradley and Dinan 2010). For example, while children who were in gestation during the Dutch famine of 1944–1945 were found to be at an increased risk of metabolic syndrome (obesity and diabetes), their risk of psychiatric disorders, such as schizophrenia, also increased (Brown et al. 2000).

Due to the complexity of the human brain, in which psychiatric disorders manifest, to say that their underlying mechanisms and aetiologies are difficult to elucidate is a gargantuan understatement. Nevertheless, significant advances have been made in the past few decades in piecing together the developmental basis of the HPA axis and psychiatric disorders, using both epidemiology and model organism approaches. The following sections will address what we currently know about stress, its underlying mechanisms, how stress may be dysregulated in disease, and the crucial relevance of stress and the HPA axis to the periconception period and disease risk.

2 What Is Stress?

Although its definition is somewhat ambiguous in the biological literature, the consensus becoming increasingly accepted is that stress entails a state of disrupted homeostasis which is counterbalanced by adaptive mechanisms known as *stress responses* (Barton 2002; Chrousos 2009). As aspects of the biotic and abiotic environment are in constant flux, organisms must continually respond to environmental changes through homeostasis mechanisms. However, excessive stimuli pose a threat to homeostasis and require a stress response to restore balance – a process also referred to as *allostasis* (Schulte 2014). Stimuli which induce a stress response may be referred to as *stressors*.

The stress response underlies the body's often extraordinary ability to respond to unexpected danger, colloquially known as the *fight or flight* response (Sorrells et al. 2009). However, stress, by definition, is something we are not adapted to cope with excessively or repeatedly. Thus, excessive environmental stress impairs an organism's fitness due to 'wear and tear' referred to as *allostatic load* (Schulte 2014). The stress response is best geared towards restoring homeostasis in the face of single or *acute* stressors, while allostatic load accumulates in the face of prolonged or *chronic* stress, which may constitute several exposures, perhaps over a substantial duration of an organism's life cycle (Sorrells et al. 2009). Arguably the most significant source of chronic stress relevant to modern life is psychological, resulting from negative socioeconomic factors such as job insecurity, financial problems, bereavement, and other personal struggles (Nargund 2015). The harsher end of chronic stress may include a prolonged state of real or perceived danger, such as domestic abuse or severe resource detriment (i.e. famine).

This chapter will adhere to the concept of stress relating to activation of the HPA axis. It is to be considered distinct from oxidative stress (excessive exposure to reactive oxygen species), which is a ubiquitous mechanism underlying several biological processes, but does not necessarily implicate the HPA axis.

2.1 *The Hypothalamic-Pituitary-Adrenal (HPA) Axis*

When the sensory systems detect a threat to homeostasis, a stress response is initiated in order to evoke endocrine and behavioural responses to enhance survival in the face of stress and ultimately restore homeostasis. Biologically, the stress response entails the initiation and regulation of a suite of endocrine pathways

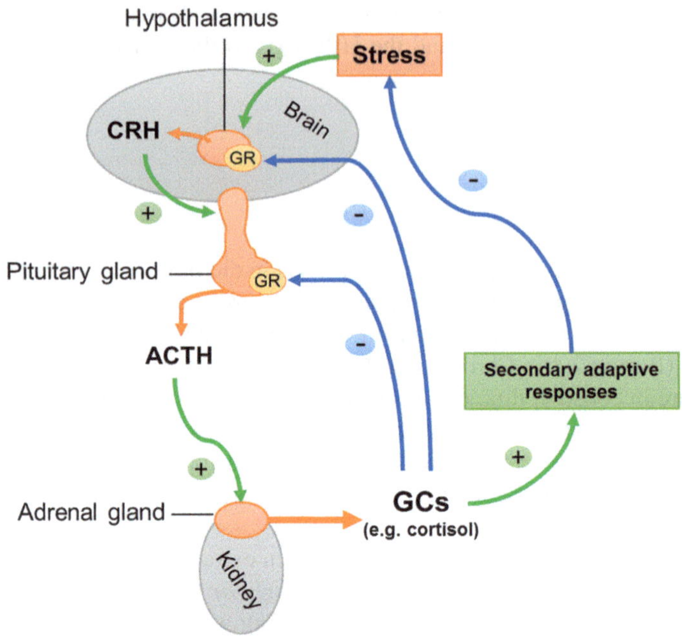

Fig. 1 Schematic diagram of the hypothalamic-pituitary-adrenal (HPA) axis. Upon registration of a stress stimulus by sensory neurones, information is relayed to the paraventricular nucleus (PVN) in the hypothalamus. The PVN continuously synthesises corticotropin-releasing factor (*CRF*) which, in response to sensory stimuli, is secreted into portal blood vessels which lead to the pituitary gland. Here, binding of CRF to its receptors induces the release of adrenocorticotropic hormone (*ACTH*), which enters the systemic circulation. Circulating ACTH reaches the adrenal cortex, situated along the perimeter of the adrenal gland, and upon reception stimulates the production and release of glucocorticoids (*GCs*) from the adrenal gland. GCs interact with glucocorticoid receptor (GR) to enact secondary adaptive responses, as well as the inhibition of the HPA axis via negative feedback. Plus signs: stimulatory effects, minus signs: inhibitory effects

embodied by the hypothalamic-pituitary-adrenal (HPA) axis (Smith and Vale 2006; Bradley and Dinan 2010) (Fig. 1). As the name would suggest, the principal structures of the HPA axis are the hypothalamus (within the brain), pituitary gland (at the base of the brain), and adrenal glands (above the kidneys). In short, registration of a stress stimulus triggers a cascade of neuronal and endocrine events, culminating in the release of glucocorticoids (GCs) as the primary stress response, which interact with glucocorticoid receptors (GRs) to enact a variety of secondary adaptive responses.

The name 'glucocorticoid' derives from early observations that the hormones are involved in glucose metabolism. The primary GC hormone in humans is cortisol, which initiates and regulates a suite of adaptive responses: it interacts with the central nervous system to induce changes in cognition and awareness; stimulates increased glucose production, providing readily available energy for responding to an immediate threat (i.e. *fight or flight*); and inhibits costly immune functions. Since the discovery of their immunosuppressive properties in the 1940s, GCs have provided useful anti-inflammatory drugs, which have been used to treat inflammatory diseases such as rheumatoid arthritis and asthma (Lupien et al. 2007). To regulate the stress response, cortisol also has an inhibitory effect on HPA activity in the hypothalamus, which establishes a negative feedback loop essential to healthy HPA axis functioning. The effects of cortisol and other GCs are mediated by the glucocorticoid receptor (GR) and a cytosolic protein complex composed of heat-shock proteins (HSPs) and expressed in almost every cell type in the body. Following stress, GCs extensively occupy GRs, which enact transcriptional modifications either via binding with transcription factors or as transcription factors themselves via direct interaction with glucocorticoid response elements (GREs). Thus, cortisol induces up- or downregulation of several genes, leading to the synthesis of enzymes responsible for glucose production, neurotrophic factors, and immunosuppressive factors. Cortisol dampens the stress response via the suppression of corticotropin-releasing factor (CRF) and adrenocorticotropic hormone (ACTH) following GR binding in the hypothalamus and pituitary gland, respectively (Fig. 1).

Abnormal HPA axis functioning is associated with numerous pathologies, including both physical and psychiatric disorders. Both genetic and environmental factors may contribute to HPA axis dysfunction, which usually implicates imbalances of GCs, GRs, or both. Within the brain, GRs occur at high concentrations in the hippocampus, which is concerned with learning, memory, and attention (Lupien et al. 2007), and in the limbic system, which is responsible for emotion (Harris et al. 2013). Therefore, imbalances in levels of GCs or GRs have the potential to adversely affect attention span, emotional state, and other aspects of cognition. The association between GCs and psychiatric disorders first became evident in the 1950s through the increased incidence of psychosis in patients receiving GC therapy. These patients displayed gradually rising euphoria or dysphoria culminating in manic episodes, a condition which became known as 'steroid psychosis' (Lupien et al. 2007). Since then, imbalances in GCs and GRs have been implicated in major depressive disorder (MDD) (Alt et al. 2010), schizophrenia (Bradley and Dinan 2010), post-traumatic stress disorder (PTSD) (Palma-Gudiel et al. 2015a, b), and almost all anxiety disorders (Faravelli et al. 2012). The development of psychiatric disorders is frequently associated with chronic stress, such as childhood trauma, suggesting that HPA axis dysregulation may be induced by prolonged allostatic load at critical developmental stages (Heim and Nemeroff 2001; Lupien et al. 2009). This is possible due to the neuroplasticity of the early brain – its ability to reorganise its structure in response to intrinsic and extrinsic stimuli (Fenoglio et al. 2006; Cramer et al. 2011) – and is now thought to be mediated by chromatin remodelling associated with GR. In rodents, for instance, repeated psychological stress leads to increased phospho-acetylation of histone H3 in the hippocampus, but this is prevented by treatment with GR antagonists (Kolber et al. 2009). Furthermore, hypomethylation of the *NR3C1* gene (encoding GR) is found in PTSD patients (Palma-Gudiel et al. 2015a, b), while long-term alterations in DNA methylation in *NR3C1* promoter regions have been suggested to mechanistically link MDD with childhood trauma (Alt et al. 2010).

Psychological illness is frequently associated with physical ill health. This may owe partly to the fact that the HPA axis does not only regulate the response to stress but also influences many other bodily processes including cardiovascular function, energy provision, fat deposition, and immune responses (Kolber et al. 2009; Sorrells et al. 2009; Bradley and Dinan 2010). Thus, as well as affecting psychological health, disruption of HPA axis function through stress may have consequences for physical health. For example, excessive production of glucose resulting from overexposure to GCs may result in metabolic disorders such as type 2 diabetes (Bradley and Dinan 2010). Furthermore, the immunosuppressive properties of GCs leave the body open to infection in states of chronic stress. In mice, for instance, chronic psychological stress and subsequent increase in endogenous GCs induce downregulation of antimicrobial peptides, increasing the severity of a bacterial skin infection (Aberg et al. 2007). In contrast, however, in some cases of acute stress, GCs may also enhance the immune response in the central nervous system (Sorrells et al. 2009).

The HPA axis comprises an ancient mechanism which is largely conserved across the vertebrate subphylum; however, some interspecies differences (and similarities) are worthy of note. Importantly, rodents utilise corticosterone instead of cortisol as their primary GC hormone. Teleost fish possess an equivalent to the HPA axis called the hypothalamic-pituitary-interrenal (HPI) axis, although the core stress response mechanism is virtually identical to its mammalian counterpart, and it is also noteworthy that fish, like humans, utilise cortisol as their principal GC hormone. In fact, some of the core endocrine components of the HPA axis are so deeply rooted in the evolutionary substrata that they form part of equivalent stress response systems in invertebrates (Ottaviani et al. 1994; Couto-Moraes et al. 2009).

Because of the conserved nature of the stress response apparatus, methods for robustly quantifying behavioural and physiological stress responses have been developed in order to study HPA axis dysregulation in model organisms, including rodents and, more recently, fish. Rodent models of chronic stress typically entail a cocktail of stressful procedures administered daily, such as physical restraint or crowding, electric shock, exposure to fox odour, constant light, or loud noises (Takahashi et al. 1998; Aberg et al. 2007; Jensen Peña et al. 2012; Howerton et al. 2013). Cortisol or corticosterone concentrations in plasma or whole body samples (in the case of fish) can be quantified using enzyme-linked immunosorbent assay (ELISA) (Cachat et al. 2010), while several behavioural paradigms have been developed to quantify anxiety-like behaviour (Fig. 2). These include the open field test, which relies on rodents' innate aversion to an unfamiliar environment and generally uses the time spent at the edge of the test arena (a behaviour called thigmotaxis) as a measure of anxiety (Prut and Belzung 2003), and the light-dark preference test, which relies on rodents' aversion to bright light (referred to as scototaxis) and uses time spent in darkness as a measure of anxiety (Bourin and Hascoët 2003; Arrant et al. 2013). Tests designed for rodents have been successfully adapted for use with zebrafish (*Danio rerio*) and other teleosts (Champagne et al. 2010; Ariyomo et al. 2013), while unique assays to measure anxiety-like behaviour in fish have also been developed, such as the novel tank diving test, which uses the depth of a fish in an unfamiliar tank as a measure of anxiety (Egan et al. 2009). For more extensive coverage of behavioural tests used to assess stress phenotypes in model organisms, readers are directed to Kumar et al. (2013) for rodents and Stewart et al. (2012) for fish.

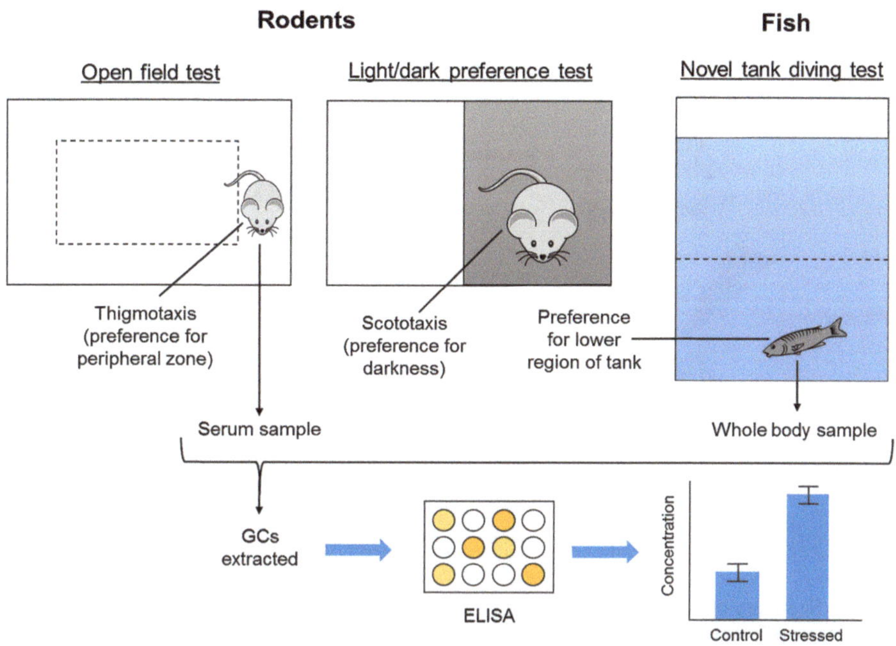

Fig. 2 Examples of quantitative behavioural and physiological stress phenotypes in rodents and fish. Strong anxiety-like behaviour (e.g. excessive thigmotaxis, scototaxis, or time spent in the lower region of a novel tank) is generally exhibited in response to HPA (mammals) or HPI (fish) axis activation by a stressor. Abnormal levels of anxiety-like behaviour detected using these measures may be indicative of dysregulation of the HPA or HPI axis, which may result from chronic stress. Following behavioural testing, cortisol may be extracted from serum or whole body samples and quantified using ELISA

3 Stress Dysregulation and Periconception

We have so far defined stress and are aware of the core components of the stress response (the HPA axis) and the pathologies that may arise from its dysregulation and how these can be translated into measurable phenotypes using model organisms. The remainder of this chapter outlines how HPA axis dysregulation, either by stress or other influences, during the periconception period may exert long-term effects on disease risk, via epigenetic alterations enacted during embryonic (or gametic) development. Pathologies associated with maternal stress, malnutrition, and alcohol exposure during pregnancy implicate HPA-axis dysregulation, which may be induced via long-term alterations to chromatin structure, in turn mediated by complex placental transduction pathways. There is also evidence to suggest that paternal stress influences embryonic development via epigenetic factors transmitted in sperm, although much remains unknown regarding the underlying mechanisms. Figure 3 presents a visual summary of molecular and phenotypic effects of periconception stress which have been identified in rodent models.

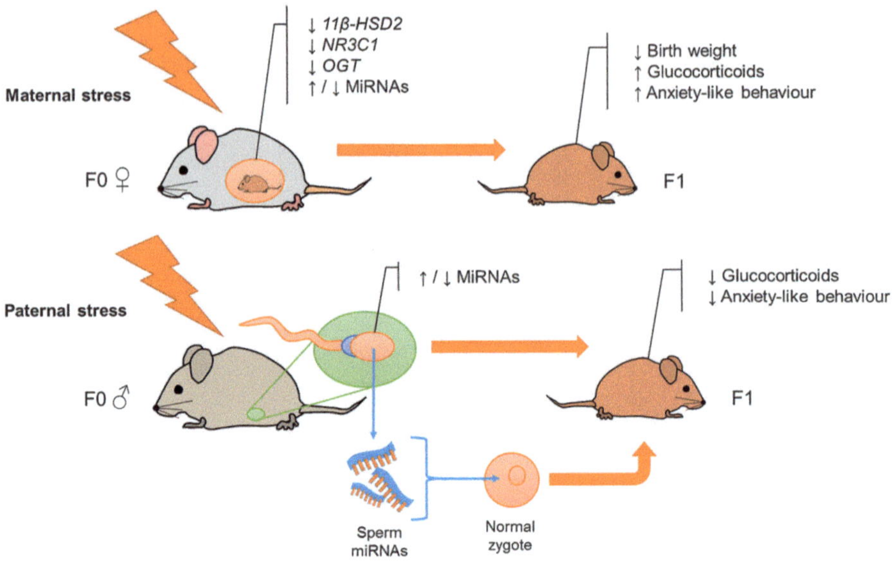

Fig. 3 Summary of molecular pathways altered by chronic stress in the periconception period and phenotypic effects in offspring. Chronic maternal stress in the prenatal period induces downregulation of *11β-HSD2* (Conradt et al. 2013), *NR3C1* (Cottrell and Seckl 2009; Conradt et al. 2013), and *OGT* (Howerton and Bale 2014) in the placenta and/or foetus and postnatal phenotypes indicative of HPA axis hyperactivity (increased glucocorticoids and anxiety-like behaviour) (Lupien et al. 2009). Chronic paternal stress in early life (Gapp et al. 2014), adolescence, or adulthood (Rodgers et al. 2013) alters sperm RNA composition and induces phenotypes indicative of suppressed HPA axis activity (decreased glucocorticoids and anxiety-like behaviour) in subsequent offspring. Insertion of sperm RNAs from stressed males into normal zygotes by microinjection recapitulates the paternal stress phenotypes in resultant pups (Gapp et al. 2014; Rodgers et al. 2015)

3.1 *Maternal Influences*

The prenatal period is now understood to be one of the most crucial stages of the human life cycle in terms of our future health and wellbeing, both physically and psychologically. Prenatal stress, which may include domestic abuse, is associated with increased risk of adverse birth outcomes, such as preterm birth (Lilliecreutz et al. 2016) and growth retardation (Cottrell and Seckl 2009), while evidence has grown over the past few decades to link psychological stress during gestation with longer-term developmental outcomes. Depression during pregnancy, which affects up to 10% of women in the UK (Vigod and Wilson 2016), with similar statistics reported in the USA (Kinsella and Monk 2009; Melville et al. 2010; Stewart 2011), has been shown to be a predictor of neurodevelopmental disorders in children and adolescents, while maternal stress during the first trimester of pregnancy is associated with increased risk of schizophrenia (Khashan et al. 2008), suggesting neurodevelopment is sensitive to stress during this early window. Prenatal famine exposure, studied in the Dutch famine cohort, has been associated with an increased risk of psychiatric disorders, including a twofold increase in schizophrenia and related conditions (Brown et al. 2000), while foetal alcohol exposure is associated with later onset of depression and anxiety (Hellemans et al. 2010). Although several factors (e.g. postnatal influences) may play a role in these observed effects, there is extensive interest in, and growing evidence for, the impact of stress on prenatal development (particularly in relation to the HPA axis) via alterations to in utero physiology and epigenetic programming (Kinsella and Monk 2009; Palma-Gudiel et al. 2015a, b). Such alterations undoubtedly involve complex interactions between the maternal environment, the placenta, and the developing embryo (Howerton et al. 2013).

GCs play several essential roles in embryonic development, particularly of the neural tissues (Harris and Seckl 2011), but overexposure to GCs resulting from stress has adverse consequences for prenatal development (Lupien et al. 2009). In rats, chronic stress during pregnancy increases corticosterone in both mother and foetus (Takahashi et al. 1998), which mediates increased anxiety-like phenotypes in adult offspring (Barbazanges et al. 1996; Lupien et al. 2009). GCs, which are employed for glucose production, are also increased in both mother and foetus during the state of chronic stress induced by undernutrition (Blondeau et al. 2001) and as a result of alcohol exposure (Liang et al. 2011). Thus, HPA axis dysregulation resulting from overexposure to GCs may underlie pathologies associated with maternal stress and undernutrition (Brown et al. 2000), as well as foetal alcohol syndrome (Hellemans et al. 2010).

The molecular aetiology of developmental programming of the HPA axis in response to prenatal stress is likely to include epigenetic alterations to target chromatin, as chromatin organisation affects the levels of expression of associated gene sequences (Cottrell and Seckl 2009). Differential expression of three key placental genes has been implicated in prenatal stress: 11β-hydroxysteroid dehydrogenase type 2 (*11β-HSD2*), glucocorticoid receptor (*NR3C1*) (Conradt et al. 2013), and O-linked N-acetylglucosamine transferase (*OGT*) (Howerton et al. 2013). In addition, a host of regulatory RNAs have been implicated in developmental programming. However, these are likely only a few of the factors contributing to the byzantine dialect between the environment, placenta, and developing embryo, in which much remains to be elucidated.

11β-HSD2 regulates foetal GC levels by converting cortisol into inert cortisone, thus protecting the foetus from GC overexposure. Maternal stress, anxiety, and undernutrition induce downregulation of *11β-HSD2*, which has been shown to correlate with reduced birth weight, as well as HPA axis dysregulation and anxiety-like behaviour (Cottrell and Seckl 2009; Conradt et al. 2013). Similar outcomes are observed in homozygous knockout mice (*11β-HSD2−/−*) (Cottrell and Seckl 2009). In rats exposed to chronic prenatal stress, foetuses possess reduced expression of *11β-HSD2* and increased CpG methylation in the *11β-HSD* promoter region in hypothalamic tissue (Jensen Peña et al. 2012), while human mothers who report anxiety during pregnancy possess greater placental methylation of *11β-HSD2* (Conradt et al. 2013). Collectively, the evidence suggests that *11β-HSD2* is an important component of the molecular interface between the maternal environment and the developing foetus and thus significant to the aetiology of stress-induced developmental pathologies.

NR3C1 is the gene encoding the glucocorticoid receptor (Conradt et al. 2013). Like GC, GRs are essential for normal development. For example, homozygous GR knockout mice die in the first few hours of life due to severely impaired lung development (Kolber et al. 2009). Likewise, reduction in *NR3C1* expression by 30–50% in transgenic mice leads to exaggerated HPA axis responses to stress (Michailidou et al. 2008). There is now evidence to link this differential expression to targeted epigenetic reprogramming in response to prenatal stress. For example, mothers who report depression during pregnancy have higher methylation of placental *NR3C1* (Conradt et al. 2013), while domestic abuse during pregnancy is significantly associated with methylation in the *NR3C1* promoter in adolescent offspring (Radtke et al. 2011). A recent meta-analysis of human DNA methylation data from 977 individuals revealed that methylation of a single CpG site in the promoter region of *NR3C1* was significantly associated with prenatal stress (Palma-Gudiel et al. 2015a).

Another factor recently implicated in the link between prenatal stress and disease risk is O-linked N-acetylglucosamine (O-GlcNAc) transferase (Ogt). The enzyme is a key cellular regulator which modifies, by addition of O-GlcNAc, protein targets responsible for chromatin remodelling (e.g. RNA polymerases, histone deacetylases) (Howerton et al. 2013). Ogt also preferentially associates with TET proteins (regulators of DNA methylation state) in close proximity to CpG-rich transcription start sites (Vella et al. 2013). Maternal stress leads to reduced expression of placental *OGT*, and *OGT*-knockout mice develop HPA axis dysregulation characteristic of that induced by stress in early pregnancy (Howerton and Bale 2014). Deficiency of Ogt is hypothesised to underlie observations of male-biased risk of neurodevelopmental disorders, as it escapes X chromosome inactivation in the placenta and is thus expressed at higher levels in females (Howerton et al. 2013). Furthermore, because O-GlcNAc is produced from glucose, Ogt is a potent sensor of cellular nutritional status and is thought to be similarly responsive to other aspects of the environment (Zachara and Hart 2004; Love and Hanover 2005; Vella et al. 2013). Because of this, and because of its interaction with TET proteins and other factors associated with chromatin remodelling (Vella et al. 2013; Howerton et al. 2013), it is plausible that Ogt is a key mediator of stress-induced epigenetic alterations associated with *11β-HSD* and *NR3C1*.

In addition to, and very likely in conjunction with, DNA methylation, small noncoding RNAs are now believed to be essential regulators at the crossroads of genes, development, and environment. MicroRNAs (miRNAs) are small noncoding RNA molecules (~22 nucleotides) which modulate gene expression by either repressing translation or inducing degradation of target mRNAs (Hollins and Cairns 2016). They are abundant in the brain and exhibit brain region-specific expression patterns in response to acute and chronic stress in animal models (Hollins and Cairns 2016), suggesting they are important in neuroplasticity. Subsequently, there is now evidence that miRNAs are key mediators of stress-induced neurodevelopmental pathologies. In response to gestational stress, one study revealed that the brains of new-born mice exhibit differential expression of over 336 miRNAs (Zucchi et al. 2013). Several of these miRNAs are involved in neurodevelopment and have been implicated in psychiatric disorders, including miR-219, which is upregulated in patients with schizophrenia. This differential miRNA expression was subsequently demonstrated to persist into the F2 generation, suggesting miRNAs may play a role in transgenerational programming of the oocyte (Yao et al. 2014) and thus may mediate epigenetic inheritance of disease risk. Interestingly, among the downregulated miRNAs were miR-200b, which is implicated in uterine contractibility, and thus may provide a putative mechanistic explanation for preterm birth associated with gestational stress (Yao et al. 2014).

When considering long-term implications of prenatal stress on HPA axis development, the neuroplasticity of the early postnatal brain (Cramer et al. 2011) must also be considered, as some lines of evidence suggest that alterations to HPA axis development in the prenatal period can be attenuated by intervention in the neonatal period. For example, rats exposed to handling during the preweaning period exhibit permanent reductions in corticosterone secretion and GR expression (Welberg and Seckl 2008), and consequently, neonatal handling has been found to eliminate some of the adverse effects of foetal alcohol exposure, such as increased weight gain (Weinberg et al. 1995), and HPA axis hyperactivity (Ogilvie and Rivier 1997). However, subsequent experiments have produced conflicting results in this regard (Gabriel et al. 2000).

3.2 *Paternal Influences*

The vast majority of literature on parental environmental influences on HPA axis development has focused on maternally mediated effects. Understandably, given that humans are confined to the maternal environment for the first 9 months, it was long thought that the paternal environment was of little importance. However, it has since become apparent that the spermatozoon provides to the embryo more than simply a haploid genome, and subsequently the paternal environment (particularly paternal stress) is becoming increasingly implicated in offspring disease risk, including HPA axis dysregulation.

Chronic psychological stress has long been perceived to be a potential risk factor in male infertility, and, although epidemiological studies have produced conflicting conclusions regarding the association, evidence is building that chronic psychological stress can significantly impair aspects of male fertility (Nargund 2015). Several clinical studies have now demonstrated an inverse relationship between psychological stress and semen parameters. For example, a recent analysis revealed an association between perceived stress or recent stressful life events and a reduction in sperm concentration, motility, and normal morphology (Janevic et al. 2014). Mediated by GCs, stimulation of the HPA axis is now believed to have a direct inhibitory effect on the hypothalamic-pituitary-gonadal (HPG) axis, which drives key reproductive functions in both sexes, including spermatogenesis (Nargund 2015). Specifically, GCs inhibit the release of gonadotropin-releasing hormone (GnRH) from the hypothalamus, the downstream consequences of which include a reduction in testosterone, which is an essential regulator of spermatogenesis at several stages (Smith and Walker 2014; Nargund 2015).

In contrast to maternal effects, in which germ line-mediated effects are difficult to discern from in utero effects on development, paternal effects on phenotype are more likely to represent the germ line transmission of environmental information. Germ line epigenetic inheritance has long been a puzzle due to the problem of erasure: the DNA methylation status of the parental genomes is reset during the first few cell divisions, and thus it is widely thought that most alterations to methylation acquired during the parents' lifetime are erased (Cantone and Fisher 2013). However, acquired methylation changes may escape erasure in some cases, and other types of transmissible epigenetic factors are not subject to erasure, including a host of regulatory RNAs. Specifically, miRNAs have been heavily implicated, while much is yet to be deciphered regarding the possible role of sperm histones and protamines in the conveyance of environmental information.

A handful of studies have identified heritable alterations in measurable phenotypic aspects of HPA axis activity induced by paternal stress at different developmental stages. Male mice subjected to a chronic stress paradigm (maternal separation) in early life develop depression-like symptoms, as well as phenotypes consistent with dampened HPA axis responsivity such as reduced anxiety-like behaviour and reduced corticosterone in response to stress. These phenotypes were found to be inherited by the offspring, even when fertilisation was carried out in vitro. RNAs were found to be integral to this environmental inheritance, as the injection of sperm RNA from traumatised males into normal zygotes recapitulated the observed phenotypes in the resulting pups (Gapp et al. 2014).

In addition to stress in early life, chronic stress experienced both in adolescence and during spermatogenesis (approx. 42 days in mice) in adulthood has been found to induce heritable alterations in measurable aspects of HPA axis activity. One dramatic example is the inheritance of a Pavlovian response, in which adult male mice were conditioned to associate the odour of acetophenone with an electric shock. The offspring of these mice, when presented with the same odour, exhibited a startle response without ever experiencing the electric shock. *Olfr151*, the gene encoding the odorant receptor for acetophenone, was found to possess CpG hypomethylation in the sperm of both F0-conditioned and F1-naïve males (Dias and Ressler 2014). However, whether the methylation state escaped erasure or was inherited by another mechanism is not clear.

A similar study (Rodgers et al. 2013) reported that male mice subjected to a 42-day chronic stress paradigm in either adolescence or adulthood sired offspring with dampened HPA axis activity, characterised by significantly lower corticosterone in response to stress compared to controls. These offspring also exhibited altered transcriptional profiles in the hypothalamus, including enriched expression of GC-responsive genes and gene sets associated with chromatin remodelling (e.g. histone acetyltransferases). The researchers identified nine miRNAs exclusively expressed in the sperm of stressed males, the predicted targets of which included DNA methyltransferase 3a (*DNMT3a*), a critical regulator of de novo DNA methylation (Rodgers et al. 2013). Remarkably, in a subsequent study, these authors reported that inserting only these nine miRNAs into normal zygotes was sufficient to induce the same phenotype indicative of paternal stress (Rodgers et al. 2015). Similarly, it has been reported that injection of candidate miRNAs into normal zygotes recapitulates hereditary metabolic syndrome associated with paternal obesity (Grandjean et al. 2015). Thus, taken together, the evidence provides a strong case for miRNAs as a principal language of environmental inheritance.

Another interesting observation derived by comparing the published experiments is that similar hereditary HPA axis dysregulation occurs in response to paternal stress, irrespective whether stress is experienced in early life (Gapp et al. 2014), adolescence (Rodgers et al. 2013), or adulthood (Rodgers et al. 2013; Dias and Ressler 2014). This suggests that even though the phenotypes were induced in response to stress in different developmental stages, the underlying mechanism may be very similar if not the same. Extracellular vesicles are hypothesised to be important for intercellular communication via the exchange of genetic information in plants and animals (Mittelbrunn and Sánchez-Madrid 2012) and may be responsible for the transport of stress-induced miRNAs into sperm. It is also possible that stress-induced testosterone deficiency (Nargund 2015) may play a role in miRNA-mediated inheritance, as testosterone is known to regulate the expression of several miRNAs in Sertoli cells in the testes (Panneerdoss et al. 2012; Smith and Walker 2014). Interestingly, two of the nine stress-responsive sperm miRNAs discovered by Rodgers et al. (2013) (miR-25c and miR-375) are also regulated by testosterone, as shown using a mouse model of testosterone deprivation (Panneerdoss et al. 2012). MiR-375, well-characterised in terms of function, is important for the development of the pancreas and pituitary gland, while little is known about the miR-25 family except that they are implicated in cardiac function (Wahlquist et al. 2014).

Another possibility is that stress may influence offspring phenotypes via post-translational modification to sperm chromatin structure, specifically histones and protamines. Chromatin undergoes extensive reorganisation during spermatogenesis, in which most histones are supplanted by protamines (Luense et al. 2016). Numerous unique protamine modifications, particularly acetylation and methylation, have been discovered in human and mouse sperm (Brunner et al. 2014), prompting the hypothesis that these decorations may play an important role in the epigenetic regulation of embryonic development following fertilisation and furthermore may represent mediators of germ line epigenetic inheritance (Luense et al. 2016). Although most paternal histones and protamines are believed to be replaced by maternally inherited histones soon after fertilisation (Cantone and Fisher 2013), sperm histone marks retained at fertilisation have recently been reported to be essential for correct gene expression in *Xenopus* embryos (Teperek et al. 2016). There is still very little known regarding sperm histone and protamine post-translational modifications, including the extent to which they may be subject to external environmental influences, and thus more attention is needed in this area of research.

Although the underlying mechanisms remain elusive, it is clear that environmentally induced reprogramming occurs not just in the developing embryo but in developing germ cells. The observation that the same phenotypes induced by paternal stress in early life and adolescence is also induced by stress during spermatogenesis suggests that, rather than resulting from long-term alterations to germ cell precursors, modifications to maturing germ cells occur transiently in response to long-term alterations to HPA axis functionality. If this is the case, effective therapy and restoration of normal HPA axis function may halt or at least reduce the modification of maturing germ cells, preventing the inheritance of pathologies. Alternatively, if miRNAs do indeed constitute the principal language of environmental inheritance, blocking those miRNAs upregulated by paternal stress (or supplementing those downregulated) may prevent this differential expression from manifesting in pathologies in the offspring.

So far, paternal effects mediated by miRNAs have been identified only in rodent models, with some evidence of similar mechanisms existing in *Caenorhabditis elegans* (Grossniklaus et al. 2013). There is evidence that environmental exposures can influence the miRNA content of human sperm (Marczylo et al. 2012), and it has been suggested that paternal trauma or experience of violence, such as in the case of war veterans and holocaust survivors, may be paternally transmitted and influence offspring mental health (Vaage et al. 2011). However, little evidence has emerged from epidemiological studies to suggest that such paternal exposures are transmissible down the human germ line (Yehuda et al. 2001; Vaage et al. 2011), and such associations may be more likely to arise due to behavioural influences on children, rather than epigenetic transmission. Whether such mechanisms exist in distantly related vertebrates, such as fish, is not known, although non-genetic transgenerational phenomena associated with environmental stress have been observed in teleost fish (Miller et al. 2012), and miRNAs are known to play an essential role in teleost spermatogenesis (Babiak 2014). If the mechanisms of inheritance in other vertebrates are similar to those being delineated in rodents, it would hint at the evolutionary significance of miRNA-mediated environmental inheritance, and it is possible that the mechanism may hold an ancient adaptive function (Grossniklaus et al. 2013).

4 Conclusions and Future Directions

The aetiologies of psychiatric disorders remain frustratingly elusive, making efforts to devise effective treatments still difficult. However, recent studies in both humans and animal models have shown promise in uncovering the molecular basis of these conditions, including altered epigenetic states resulting from exposure in early life, gestation, or preconception. As high-throughput sequencing technologies and other molecular tools become more affordable and accessible, it will be possible to further address knowledge gaps pertaining to the mechanisms underlying long-term effects of periconception stress. For instance, although specific chromatin marks and regulatory RNAs have been implicated in long-term effects of parental stress, there remain several such entities, such as long noncoding RNAs (lncRNAs), the functions of which we know very little about (Morris and Mattick 2014).

A well-established toolset already exists for studying stress dysregulation in model organisms, which continue to help further our understanding of this complex set of processes. There is now increasing interest in non-mammalian models, specifically zebrafish, which present an increasingly attractive avenue for the exploration of periconception stress. The rapid life cycle and easily manipulated transparent embryos of the zebrafish (*D. rerio*) have already made them one of the most powerful vertebrate tools available to embryologists, and there exists a well-developed toolset for studying their behavioural and physiological stress phenotypes (Cachat et al. 2010; Stewart et al. 2012). Zebrafish may also present a unique, high-throughput model for epigenetic effects associated with spermatogenesis, the duration of which is a mere 6 days in this species (Leal et al. 2009).

In addition, having uncovered previously unknown mechanisms of environmental inheritance in model organisms, further attention may be directed to epidemiology to determine the significance of these mechanisms in human populations. Unique miRNA profiles have already been identified in the sperm of smokers versus non-smokers (Marczylo et al. 2012), suggesting that other environmental influences, particularly stress, may affect gametic chromatin, with consequences for subsequent embryos. There is therefore a need to characterise miRNAs from gametes derived from humans suffering from chronic stress, as these may provide valuable molecular markers for risk of HPA axis dysregulation in subsequent generations.

In conclusion, an improved mechanistic understanding of environmental predisposition to HPA axis-related pathologies will have major benefits to public health, in the interests of both treatment and prevention. Increased knowledge of molecular pathways underlying disease risk may provide important biomarkers, such that those already at risk of psychiatric disorders may be identified, enabling early intervention to minimise long-term suffering. Increased knowledge of disease processes may also pave the way for the development of therapeutic agents to counteract the adverse effects of parental stress on offspring disease risk. Finally, increased awareness of environmental influences on development will help to further inform human lifestyles and behaviour, such that risk to subsequent generations is minimised.

References

Aberg KM, Radek KA, Choi E-H et al (2007) Psychological stress downregulates epidermal antimicrobial peptide expression and increases severity of cutaneous infections in mice. J Clin Invest 117:3339–3349. doi:10.1172/JCI31726

Alt SR, Turner JD, Klok MD et al (2010) Differential expression of glucocorticoid receptor transcripts in major depressive disorder is not epigenetically programmed. Psychoneuroendocrinology 35:544–556. doi:10.1016/j.psyneuen.2009.09.001

Ariyomo TO, Carter M, Watt PJ (2013) Heritability of boldness and aggressiveness in the zebrafish. Behav Genet 43:161–167. doi:10.1007/s10519-013-9585-y

Arrant AE, Schramm-Sapyta NL, Kuhn CM (2013) Use of the light/dark test for anxiety in adult and adolescent male rats. Behav Brain Res 256:119–127. doi:10.1016/j.bbr.2013.05.035

Babiak I (2014) MicroRNA in teleost fish. Genome Biol Evol 6:1911–1937. doi:10.1093/gbe/evu151

Barbazanges A, Piazza PV, Le Moal M, Maccari S (1996) Maternal glucocorticoid secretion mediates long-term effects of prenatal stress. J Neurosci 16:3943–3949

Barton BA (2002) Stress in fishes: a diversity of responses with particular reference to changes in circulating corticosteroids. Integr Comp Biol 42:517–525. doi:10.1093/icb/42.3.517

Blondeau B, Lesage J, Czernichow P et al (2001) Glucocorticoids impair fetal beta-cell development in rats. Am J Physiol Endocrinol Metab 281:E592–E599. doi:10.1016/0024-3205(90)90607-s

Bourin M, Hascoët M (2003) The mouse light/dark box test. Eur J Pharmacol 463:55–65. doi:10.1016/S0014-2999(03)01274-3

Bradley AJ, Dinan TG (2010) A systematic review of hypothalamic-pituitary-adrenal axis function in schizophrenia: implications for mortality. J Psychopharmacol 24:91–118. doi:10.1177/1359786810385491

Brown AS, Van Os J, Driessens C et al (2000) Further evidence of relation between prenatal famine and major affective disorder. Am J Psychiatry 157:190–195. doi:10.1176/appi.ajp.157.2.190

Brunner AM, Nanni P, Mansuy IM (2014) Epigenetic marking of sperm by post-translational modification of histones and protamines. Epigenetics Chromatin 7:2. doi:10.1186/1756-8935-7-2

Cachat J, Stewart A, Grossman L et al (2010) Measuring behavioral and endocrine responses to novelty stress in adult zebrafish. Nat Protoc 5:1786–1799. doi:10.1038/nprot.2010.140

Cantone I, Fisher AG (2013) Epigenetic programming and reprogramming during development. Nat Struct Mol Biol 20:282–289. doi:10.1038/nsmb.2489

Champagne DL, Hoefnagels CCM, de Kloet RE, Richardson MK (2010) Translating rodent behavioral repertoire to zebrafish (Danio rerio): relevance for stress research. Behav Brain Res 214:332–342. doi:10.1016/j.bbr.2010.06.001

Chrousos GP (2009) Stress and disorders of the stress system. Nat Rev Endocrinol 5:374–381

Conradt E, Lester BM, Appleton AA et al (2013) The roles of DNA methylation of NR3C1 and 11β-HSD2 and exposure to maternal mood disorder in utero on newborn neurobehavior. Epigenetics 8:1321–1329. doi:10.4161/epi.26634

Cottrell EC, Seckl JR (2009) Prenatal stress, glucocorticoids and the programming of adult disease. Front Behav Neurosci 3:19. doi:10.3389/neuro.08.019.2009

Couto-Moraes R, Palermo-Neto J, Markus RP (2009) The immune-pineal axis. Ann N Y Acad Sci 1153:193–202. doi:10.1111/j.1749-6632.2008.03978.x

Cramer SC, Sur M, Dobkin BH et al (2011) Harnessing neuroplasticity for clinical applications. Brain 134:1591–1609

Dias BG, Ressler KJ (2014) Parental olfactory experience influences behavior and neural structure in subsequent generations. Nat Neurosci 17:89–96. doi:10.1038/nn.3594

Egan RJ, Bergner CL, Hart PC et al (2009) Understanding behavioral and physiological phenotypes of stress and anxiety in zebrafish. Behav Brain Res 205:38–44. doi:10.1016/j.bbr.2009.06.022

Faravelli C, Lo Sauro C, Lelli L et al (2012) The role of life events and HPA axis in anxiety disorders: a review. Curr Pharm Des 18:5663–5674

Fenoglio KA, Brunson KL, Baram TZ (2006) Hippocampal neuroplasticity induced by early-life stress: functional and molecular aspects. Front Neuroendocrinol 27:180–192. doi:10.1016/j.yfrne.2006.02.001

Gabriel KI, Yu W, Ellis L, Weinberg J (2000) Postnatal handling does not attenuate hypothalamic-pituitary-adrenal hyperresponsiveness after prenatal ethanol exposure. Alcohol Clin Exp Res 24:1566–1574. doi:10.1111/j.1530-0277.2000.tb04576.x

Gapp K, Jawaid A, Sarkies P et al (2014) Implication of sperm RNAs in transgenerational inheritance of the effects of early trauma in mice. Nat Neurosci 17:667–669. doi:10.1038/nn.3695

Grandjean V, Fourré S, De Abreu DAF et al (2015) RNA-mediated paternal heredity of diet-induced obesity and metabolic disorders. Sci Rep 5:18193. doi:10.1038/srep18193

Grossniklaus U, Kelly B, Ferguson-Smith AC et al (2013) Transgenerational epigenetic inheritance: how important is it? Nat Rev Genet 14:228–235. doi:10.1038/nrg3435

Harris A, Seckl J (2011) Glucocorticoids, prenatal stress and the programming of disease. Horm Behav 59:279–289. doi:10.1016/j.yhbeh.2010.06.007

Harris AP, Holmes MC, De Kloet ER et al (2013) Mineralocorticoid and glucocorticoid receptor balance in control of HPA axis and behaviour. Psychoneuroendocrinology 38:648–658. doi:10.1016/j.psyneuen.2012.08.007

Heim C, Nemeroff CB (2001) The role of childhood trauma in the neurobiology of mood and anxiety disorders: preclinical and clinical studies. Biol Psychiatry 49:1023–1039. doi:10.1016/S0006-3223(01)01157-X

Hellemans KGC, Sliwowska JH, Verma P, Weinberg J (2010) Prenatal alcohol exposure: fetal programming and later life vulnerability to stress, depression and anxiety disorders. Neurosci Biobehav Rev 34:791–807. doi:10.1016/j.neubiorev.2009.06.004

Hollins SL, Cairns MJ (2016) MicroRNA: small RNA mediators of the brains genomic response to environmental stress. Prog Neurobiol 143:61–81. doi:10.1016/j.pneurobio.2016.06.005

Howerton CL, Bale TL (2014) Targeted placental deletion of OGT recapitulates the prenatal stress phenotype including hypothalamic mitochondrial dysfunction. Proc Natl Acad Sci 111:9639–9644. doi:10.1073/pnas.1401203111

Howerton CL, Morgan CP, Fischer DB, Bale TL (2013) O-GlcNAc transferase (OGT) as a placental biomarker of maternal stress and reprogramming of CNS gene transcription in development. Proc Natl Acad Sci U S A 110:5169–5174. doi:10.1073/pnas.1300065110

Janevic T, Kahn LG, Landsbergis P et al (2014) Effects of work and life stress on semen quality. Fertil Steril 102:530–538. doi:10.1016/j.fertnstert.2014.04.021

Jensen Peña C, Monk C, Champagne FA et al (2012) Epigenetic effects of prenatal stress on 11β-hydroxysteroid dehydrogenase-2 in the placenta and fetal brain. PLoS One 7:e39791. doi:10.1371/journal.pone.0039791

Kalueff A V, Stewart AM, Gerlai R (2014) Zebrafish as an emerging model for studying complex brain disorders. Trends Pharmacol Sci 35:63–75. doi:10.1016/j.tips.2013.12.002

Khashan AS, Abel KM, McNamee R et al (2008) Higher risk of offspring schizophrenia following antenatal maternal exposure to severe adverse life events. Arch Gen Psychiatry 65:146. doi:10.1001/archgenpsychiatry.2007.20

Kinsella MT, Monk C (2009) Impact of maternal stress, depression & anxiety on fetal neurobehavioral development. Clin Obstet Gynecol 52:425–440. doi:10.1097/GRF.0b013e3181b52df1. Impact

Kolber BJ, Wieczorek L, Muglia LJ (2009) HPA axis dysregulation and behavioral analysis of mouse mutants with altered GR and MR function. Stress 11:321–338. doi:10.1080/10253890701821081.HPA

Kumar V, Bhat ZA, Kumar D (2013) Animal models of anxiety: a comprehensive review. J Pharmacol Toxicol Methods 68:175–183. doi:10.1016/j.vascn.2013.05.003

Leal MC, Cardoso ER, Nóbrega RH et al (2009) Histological and stereological evaluation of zebrafish (Danio rerio) spermatogenesis with an emphasis on spermatogonial generations. Biol Reprod 81:177–187. doi:10.1095/biolreprod.109.076299

Liang G, Chen M, Pan X et al (2011) Ethanol-induced inhibition of fetal hypothalamic–pituitary–adrenal axis due to prenatal overexposure to maternal glucocorticoid in mice. Exp Toxicol Pathol 63:607–611. doi:10.1016/j.etp.2010.04.015

Lilliecreutz C, Larén J, Sydsjö G, Josefsson A (2016) Effect of maternal stress during pregnancy on the risk for preterm birth. BMC Pregnancy Childbirth 16:1–8. doi:10.1186/s12884-015-0775-x

Love DC, Hanover JA (2005) The hexosamine signaling pathway: deciphering the "O-GlcNAc code". Sci STKE 2005:re13. doi: stke.3122005re13 [pii]\r10.1126/stke.3122005re13

Luense LJ, Wang X, Schon SB et al (2016) Comprehensive analysis of histone post-translational modifications in mouse and human male germ cells. Epigenetics Chromatin 9:24. doi:10.1186/s13072-016-0072-6

Lupien SJ, Maheu F, Tu M et al (2007) The effects of stress and stress hormones on human cognition: implications for the field of brain and cognition. Brain Cogn 65:209–237. doi:10.1016/j.bandc.2007.02.007

Lupien SJ, McEwen BS, Gunnar MR, Heim C (2009) Effects of stress throughout the lifespan on the brain, behaviour and cognition. Nat Rev Neurosci 10:434–445. doi:10.1038/nrn2639

Marczylo EL, Amoako AA, Konje JC et al (2012) Smoking induces differential miRNA expression in human spermatozoa: a potential transgenerational epigenetic concern? Epigenetics 7:432–439. doi:10.4161/epi.19794

Melville JL, Gavin A, Guo Y et al (2010) Depressive disorders during pregnancy: prevalence and risk factors in a large urban sample. Obstet Gynecol 116:1064–1070. doi:10.1097/AOG.0b013e3181f60b0a

Michailidou Z, Carter RN, Marshall E et al (2008) Glucocorticoid receptor haploinsufficiency causes hypertension and attenuates hypothalamic-pituitary-adrenal axis and blood pressure adaptions to high-fat diet. FASEB J 22:3896–3907. doi:10.1096/fj.08-111914

Miller GM, Watson S-A, Donelson JM et al (2012) Parental environment mediates impacts of increased carbon dioxide on a coral reef fish. Nat Clim Chang 2:858–861. doi:10.1038/nclimate1599

Mittelbrunn M, Sánchez-Madrid F (2012) Intercellular communication: diverse structures for exchange of genetic information. Nat Rev Mol Cell Biol 13:328–335. doi:10.1038/nrm3335

Morris KV, Mattick JS (2014) The rise of regulatory RNA. Nat Rev Genet 15:423–437. doi:10.1038/nrg3722

Nargund VH (2015) Effects of psychological stress on male fertility. Nat Rev Urol 12:373–382. doi:10.1038/nrurol.2015.112

Ogilvie KM, Rivier C (1997) Prenatal alcohol exposure results in hyperactivity of the hypothalamic-pituitary-adrenal axis of the offspring: modulation by fostering at birth and postnatal handling. Alcohol Clin Exp Res 21:424–429. doi:10.1111/j.1530-0277.1997.tb03786.x

Ottaviani E, Franchini A, Caselgrandi E et al (1994) Relationship between corticotropin-releasing factor and interleukin-2: evolutionary evidence. FEBS Lett 351:19–21. doi:10.1016/0014-5793(94)00802-7

Palma-Gudiel H, Córdova-Palomera A, Eixarch E et al (2015a) Maternal psychosocial stress during pregnancy alters the epigenetic signature of the glucocorticoid receptor gene promoter in their offspring: a meta-analysis. Epigenetics 10:893–902

Palma-Gudiel H, Córdova-Palomera A, Leza JC, Fañanás L (2015b) Glucocorticoid receptor gene (NR3C1) methylation processes as mediators of early adversity in stress-related disorders causality: a critical review. Neurosci Biobehav Rev 55:520–535. doi:10.1016/j.neubiorev.2015.05.016

Panneerdoss S, Chang Y-F, Buddavarapu KC et al (2012) Androgen-responsive MicroRNAs in mouse Sertoli cells. PLoS One 7:e41146. doi:10.1371/journal.pone.0041146

Prut L, Belzung C (2003) The open field as a paradigm to measure the effects of drugs on anxiety-like behaviors: a review. Eur J Pharmacol 463:3–33. doi:10.1016/S0014-2999(03)01272-X

Radtke KM, Ruf M, Gunter HM et al (2011) Transgenerational impact of intimate partner violence on methylation in the promoter of the glucocorticoid receptor. Transl Psychiatry 1:e21. doi:10.1038/tp.2011.21

Rodgers AB, Morgan CP, Bronson SL et al (2013) Paternal stress exposure alters sperm MicroRNA content and reprograms offspring HPA stress axis regulation. J Neurosci 33:9003–9012. doi:10.1523/JNEUROSCI.0914-13.2013

Rodgers AB, Morgan CP, Leu NA, Bale TL (2015) Transgenerational epigenetic programming via sperm microRNA recapitulates effects of paternal stress. Proc Natl Acad Sci U S A 112: 13699–13704. doi:10.1073/pnas.1508347112. 2015:1–6

Schulte PM (2014) What is environmental stress? Insights from fish living in a variable environment. J Exp Biol 217:23–34. doi:10.1242/jeb.089722

Smith LB, Walker WH (2014) The regulation of spermatogenesis by androgens. Semin Cell Dev Biol 30:2–13

Smith S, Vale W (2006) The role of the hypothalamic-pituitary-adrenal axis in neuroendocrine responses to stress. Dialogues Clin Neurosci 8:383

Sorrells SF, Caso JR, Munhoz CD, Sapolsky RM (2009) The stressed CNS: when glucocorticoids aggravate inflammation. Neuron 64:33–39

Stewart DE (2011) Depression during pregnancy. N Engl J Med 365:1605–1611. doi:10.1056/NEJMcp1102730

Stewart A, Gaikwad S, Kyzar E et al (2012) Modeling anxiety using adult zebrafish: a conceptual review. Neuropharmacology 62:135–143. doi:10.1016/j.neuropharm.2011.07.037

Takahashi LK, Turner JG, Kalin NH (1998) Prolonged stress-induced elevation in plasma corticosterone during pregnancy in the rat: implications for prenatal stress studies. Psychoneuroendocrinology 23:571–581. doi:10.1016/S0306-4530(98)00024-9

Teperek M, Simeone A, Gaggioli V et al (2016) Sperm is epigenetically programmed to regulate gene transcription in embryos. Genome Res 26:1034–1046. doi:10.1101/gr.201541.115

Vaage AB, Thomsen PH, Rousseau C et al (2011) Paternal predictors of the mental health of children of Vietnamese refugees. Child Adolesc Psychiatry Ment Health 5:2. doi:10.1186/1753-2000-5-2

Vella P, Scelfo A, Jammula S et al (2013) Tet proteins connect the O-linked N-acetylglucosamine transferase Ogt to chromatin in embryonic stem cells. Mol Cell 49:645–656. doi:10.1016/j.molcel.2012.12.019

Vigod SN, Wilson CA (2016) Depression in pregnancy. BMJ 352:i1547. doi:10.1136/bmj.i1547

Vos T, Barber RM, Bell B et al (2015) Global, regional, and national incidence, prevalence, and years lived with disability for 301 acute and chronic diseases and injuries in 188 countries, 1990-2013: a systematic analysis for the global burden of disease study 2013. Lancet London England 386:743–800. doi:10.1016/S0140-6736(15)60692-4

Wahlquist C, Jeong D, Rojas-Muñoz A et al (2014) Inhibition of miR-25 improves cardiac contractility in the failing heart. Nature 508:531–535. doi:10.1038/nature13073

Weinberg J, Kim CK, Yu W (1995) Early handling can attenuate adverse effects of fetal ethanol exposure. Alcohol 12:317–327. doi:10.1016/0741-8329(95)00005-C

Welberg LAM, Seckl JR (2008) Prenatal stress, glucocorticoids and the programming of the brain. J Neuroendocrinol 13:113–128. doi:10.1111/j.1365-2826.2001.00601.x

Yao Y, Robinson AM, Zucchi FC et al (2014) Ancestral exposure to stress epigenetically programs preterm birth risk and adverse maternal and newborn outcomes. BMC Med 12:121. doi:10.1186/s12916-014-0121-6

Yehuda R, Halligan SL, Bierer LM (2001) Relationship of parental trauma exposure and PTSD to PTSD, depressive and anxiety disorders in offspring. J Psychiatr Res 35:261–270. doi:10.1016/S0022-3956(01)00032-2

Zachara NE, Hart GW (2004) O-GlcNAc a sensor of cellular state: the role of nucleocytoplasmic glycosylation in modulating cellular function in response to nutrition and stress. Biochim Biophys Acta Gen Subj 1673:13–28. doi:10.1016/j.bbagen.2004.03.016

Zucchi FCR, Yao Y, Ward ID et al (2013) Maternal stress induces epigenetic signatures of psychiatric and neurological diseases in the offspring. PLoS One 8:e56967. doi:10.1371/journal.pone.0056967

The Long-Term Effects of the Periconceptional Period on Embryo Epigenetic Profile and Phenotype; The Paternal Role and His Contribution, and How Males Can Affect Offspring's Phenotype/Epigenetic Profile

Emma S. Lucas and Adam J. Watkins

Abstract The number of adults afflicted with heart disease, obesity and diabetes, central components of metabolic disorder, has grown rapidly in recent decades, affecting up to one quarter of the world's population. Typically, these diseases are attributed to lifestyle factors such as poor diet, lack of exercise and smoking. However, studies have now identified strong associations between patterns of growth during foetal and neonatal life and an increase predisposition towards developing heart disease, obesity and diabetes in adult life. While the connection between a mother's diet and the long-term health of her offspring has been studied in great detail, our understanding of whether offspring health might be affected by a father's diet remains limited. Greater insight into the impact that paternal nutrition has on sperm quality, epigenetic status and potential offspring programming mechanisms is needed to redress this parental-programming knowledge imbalance. Disturbances in paternal reproductive epigenetic status represents one key mechanism linking paternal diet with the programing of offspring development and adult health, as many enzymatic processes involved in epigenetic regulation use metabolic intermediates to modify DNA and histones. Here, poor paternal nutrition could result in perturbed sperm and testicular epigenetic status, impacting on post-fertilisation gene transcriptional regulation and protein expression in offspring tissues, resulting in increased incidences of metabolic disorder in adult life.

Keywords Paternal nutrition • DNA • Histones • Metabolic syndrome • Reproductive fitness

E.S. Lucas
Division of Reproductive Health, Clinical Science Research Laboratories, Warwick Medical School, University of Warwick, Coventry CV2 2DX, UK

A.J. Watkins (✉)
Aston Research Centre for Healthy Ageing, School of Life and Health Sciences, Aston University, Birmingham, B4 7ET, UK
e-mail: a.watkins1@aston.ac.uk

A. Fazeli, W.V. Holt (eds.), *Periconception in Physiology and Medicine*, Advances in Experimental Medicine and Biology 1014,
DOI 10.1007/978-3-319-62414-3_8

1 Introduction

It is estimated that up to 60% of global mortality is attributed to non-communicable diseases such as the metabolic syndrome (Wang et al. 2011). Typically, such diseases have been attributed to adult lifestyle characteristics such as poor diet, sedentary behaviour and smoking. However, research from world-wide human cohort data and a diverse range of animal models have established clear links between the size of individuals at birth, altered growth during infancy and an increased risk of developing cardiovascular disease, type 2 diabetes and obesity in adult life. Similarly, adult-onset degenerative diseases such as hypertension, osteoporosis and obstructive airway disease have also been linked to perturbed patterns of growth during foetal development and early life.

Studies have demonstrated that deficiencies or excess in a range of maternal macro- and micronutrients impact negatively on reproductive fitness, embryonic and foetal development, offspring growth and long-term health. Animal model studies, in conjunction with retrospective analyses of human cohort data, have shown that all stages of foetal and embryonic development, as well as the development and maturation of sperm and oocytes, are influenced directly by suboptimal environmental conditions, whether in vivo (parental diet) or in vitro (embryo culture) (Lee 2015; Yeung and Druschel 2013). Historically, studies have focused on the impact of maternal dietary deficiencies on long-term offspring growth. These have centred on the use of diets low in protein (Fleming et al. 2015), reduced in caloric content (Reynolds et al. 2015) or deficient for specific micronutrients such as iron (Radlowski and Johnson 2013). However, as the world today is faced with the dual burden of both excessive under- and overnutrition in human populations (Kelly et al. 2008; Martin et al. 2010b), a new focus into the impact of energy-dense diets on parental reproductive fitness and offspring long-term health has emerged.

While the significance of optimal maternal nutritional status during specific periods of gamete maturation and post-fertilisation development has received detailed investigation, it has only been over the past several years that research has begun to focus on the importance of male physiology and nutritional status at the time of conception. In men, obesity impacts negatively on testosterone levels, sperm number, sperm motility, embryonic development and live birth rates (Bakos et al. 2011a; Chavarro et al. 2011a; Kort et al. 2006b). In animal models, paternal nutritional manipulation has been shown to impact on embryonic metabolism, foetal growth and skeletal formation as well as long-term cardiovascular and metabolic health (Binder et al. 2012; Carone et al. 2010; Lambrot et al. 2013; Ng et al. 2010; Watkins and Sinclair 2014). Indeed, our increased understanding of potential mechanisms through which impaired paternal health may programme offspring long-term health has resulted in a new interest in defining semen quality and assessing sperm epigenetic status.

Global reproductive fitness is not only impacted on by the burden of increased levels of energy dense diets of low micronutrient content, but is faced also with the increase used of Assisted Reproductive Techniques (ART) such as IVF. The potential long-term impacts of routine human embryo culture and manipulations were brought into question following observations that the in vitro manipulation, culture and transfer of cattle, sheep and mouse embryos results in significant changes in offspring growth, rates of neonatal mortality, blood pressure and gluconeogenic regulation and life span (Wrenzycki et al. 2004; Watkins et al. 2007; Rexhaj et al. 2013; Grace and Sinclair 2009; Calle et al. 2012). As a consequence, retrospective analyses of human children conceived through ART procedures have identified some associations between human embryo in vitro manipulation and low birth weight, pre-term delivery, altered epigenetic status, elevated blood pressure and glucose intolerance (Yeung and Druschel 2013; Scherrer et al. 2012; Nelissen et al. 2013; Dumoulin et al. 2010; Ceelen et al. 2008). With infertility affecting between 8% and 12% of reproductive-aged couples globally, and an estimated 1–3% of all births arising from some form of ART in developed countries (Inhorn and Patrizio 2015; Ombelet et al. 2008), more detailed and mechanistic studies are required to draw definitive conclusions regarding potential long-term health impacts of parental physiology, gamete quality and human embryo culture and manipulation on long-term health of resultant children.

With these topics in mind, the aim of this chapter is to provide insight and contemporary overview of mechanisms linking paternal environmental and nutritional status with early developmental processes and ultimately adult cardiovascular, metabolic and ageing health. Initially, we will summarise briefly current understanding of the impact of maternal nutrition during the periconceptional period, before assessing the role of paternal nutrition on male reproductive fitness and offspring development. This chapter will detail the impact of parental diet and physiological status on gamete quality, embryo development, foetal growth and long-term offspring health. The use of animal models has become a key tool in understanding the underlying mechanisms linking environmental factors with the programming of offspring health. Rodent models form a large proportion of studies due to their quick generation times, well characterised genomes and similar patterns of early embryo development to humans. However, larger animal models afford the opportunity to study foetal development over a longer period of time and, depending on strain, often only carry singleton foetuses (more reflective of human gestation). Therefore, this chapter will refer to data from human, rodent and, where appropriate, livestock species studies.

2 Developmental Programming

The Developmental Origins of Health and Disease (DOHaD) hypothesis proposes that the maternal reproductive and uterine environment experienced during gestation can impact on offspring non-communicable disease risk susceptibility in adult life. Human clinical and epidemiological studies have accumulated a wealth of correlative data linking the quantity and/or quality of maternal gestational nutrition, foetal and placental development, birth weight, neonatal growth and chronic disease risk in later life. These studies reproducibly demonstrate significant associations between altered birth weight (both increased and decreased) and the development of cardiovascular dysfunction, metabolic syndrome, psychological disorders and degenerative conditions such as osteoporosis (Barker et al. 1993; Hanson and Gluckman 2014). The first direct evidence that an adverse intrauterine environment may impact on long-term health came from the analysis of middle aged men and women in a Hertforshire cohort (Barker et al. 1993). Here, individuals who had a low weight at birth had increased death rates from coronary heart disease in adult life. Similar associations have been observed in separate human cohorts from around the world (Cheung et al. 2000; Dabelea et al. 1999; Forsen et al. 1997; Rich-Edwards et al. 1997).

From these diverse human epidemiological studies came the proposition that maternal nutritional status during gestation induces physiological changes in foetal development and growth, resulting in significant long-term changes in adult health. Nutritional signals received by the developing foetus promote adaptive responses within developing tissues and organism to maximise foetal and neonatal development for a perceived postnatal nutritional environment. However, if the pre- and postnatal nutritional environments are mismatched, these adaptations can become inappropriate and maladaptive for postnatal and adult life, resulting in increased risk for adult ill-health. With the link established between maternal diet, gamete quality and post-fertilisation development, it is easy then to apply these observations and mechanistic insights onto paternally mediated programming effects.

3 Maternal Nutrition

3.1 Obesity and Overnutrition

In the UK, between 1989 and 2007, the rates of maternal obesity have increased from 7.6% to 15.6% 9 (Heslehurst et al. 2010), while in the USA, reports indicate that the prevalence of pre-pregnancy obesity has risen to 28.5% (Hinkle et al. 2012). The negative impacts of obesity on female reproduction are well established; studies have shown that oocyte quality in obese women can be affected by elevated levels of triglycerides and free fatty acids within the follicular fluid surrounding the oocyte (Robker et al. 2009). Increases in insulin and androgen levels, ovarian dysfunction, type 2 diabetes, polycystic ovarian syndrome (PCOS) and endometrial hyperplasia are also observed as direct effects of maternal obesity (Kulie et al. 2011).

Mirroring observation in women, feeding female mice a high fat diet results in increased ovarian lipid accumulation, elevated levels of apoptosis within the ovary, reduced fertilisation rates and altered embryo mitochondrial reactive oxygen species (ROS) generation (Igosheva et al. 2010; Jungheim et al. 2010). Maternal high fat diet also perturbs mitochondrial metabolism, altering DNA content and membrane potential, inducing endoplasmic reticulum stress and caspase activation

(Igosheva et al. 2010; Jungheim et al. 2010; Wu et al. 2011). Those embryos that do develop form smaller foetuses which then display postnatal 'catch-up-growth', a well-established marker for adverse adult health in human epidemiological studies (Barker et al. 1993).

3.2 Nutrient-Deficient Diets

Epidemiological observations in adult offspring of women exposed to the Dutch Hunger Winter famine (1944–1945) have provided an opportunity to assess the long-term impact of maternal undernutrition restricted to specific period or stages of pregnancy. Maternal nutrient restriction during the first trimester of pregnancy was linked to increased prevalence of coronary heart disease, raised lipids and obesity in offspring (Ravelli et al. 1999; Roseboom et al. 2000; Roseboom et al. 2001), whereas famine occurring during late gestation led to decreased glucose tolerance in adult life (Ravelli et al. 1998).

As with maternal overnutrition, animal models have also identified maternal undernutrition during early development as a critical factor in programming offspring ill-health. Feeding female mice a low protein diet (LPD; 9% protein) exclusively during preimplantation development (first 3.5 days of gestation) induces offspring overgrowth and elevated systolic blood pressure in addition to vascular dysfunction and elevated patterns of offspring activity within an open field test (Watkins et al. 2008, 2011). Similar to the impact of a maternal high fat diet, LPD has been shown to alter both mitochondrial localisation and membrane potential (Mitchell et al. 2009). In addition, significant changes in maternal uterine amino acid profiles, insulin and glucose levels have been observed in female mice in response to periconceptional LPD (Eckert et al. 2012). These dietary-induced changes in maternal metabolic status may therefore induce metabolic stress within the preimplantation embryo, thereby altering foetal growth and development and, ultimately, programming life-long offspring ill-health (Fleming et al. 2015).

4 Paternal Programming of Offspring Development

In light of the significant impacts maternal periconceptional nutrition has on offspring development and life-long health, increased interest is being given to exploring the influence that paternal environmental manipulation can have on the development of their offspring. As with maternal mechanisms of offspring programming, animal studies have revealed themselves to be essential for the development of human-relevant models with which the underlying biological mechanisms can be investigated. For example, in Southeast Asia and Polynesia, the chewing of betel nuts is very popular. However, individuals who do so are at increased risk of developing metabolic syndrome (Lin et al. 2008) as well as increasing the risk in their own offspring (Chen et al. 2006) irrespective of offspring betel nut consumption. Interestingly, in a mouse model in which males were exposed to betel nuts in their diet prior to mating, their offspring developed hyperglycaemia, mirroring effects observed in humans, and remained manifest for up to three generations (Boucher et al. 1994).

In many species, offspring development occurs in the physical absence of their fathers. Therefore, in order for paternal experiences to influence the development of their offspring, information must be faithfully transmitted at the point of fertilisation and stably incorporated into the offspring's genome. Sperm mediated programming through epigenetic transmission of heritable changes in chromatin structure offers one potential mechanism by which altered paternal physiology can impact on offspring development. The observation that a grandfather's prepubertal (8–12 years of age) nutritional status and growth can impact on the health and mortality risk of their grandsons, but not granddaughters, identifies the potential transgenerational impact of paternal programming (Kaati et al. 2002; Pembrey et al. 2006b). Such observations also indicate potential differential programming effects of the inherited X and Y chromosomes, dependent on the sex of the offspring inheriting them. In addition to a sperm-genomic contribution, seminal plasma constituents can also interact with maternal oviductal and uterine tissues, modulating the female reproductive environment during early embryo development, a critically sensitive period in development (Sinclair and Watkins 2013). This section will, therefore, highlight animal model and human studies demonstrating paternal programming of post-fertilisation development and discuss the role of epigenetics in mediating paternal influences.

4.1 Paternal Nutrition and Reproductive Fitness

Mirroring female reproductive fitness, male fertility is closely linked to nutrition, physiological status and body size. Daily rates of sperm production and sperm quality (i.e. sperm count and sperm motility) are decreased by undernutrition in rams (Parker and Thwaites 1972; Robinson et al. 2006), while in bulls, levels of nutrition affect testicular development and sperm production (Gauthier and Berbigier 1982; Vandemark et al. 1964). Male nutritional status can also affect testicular morphology (area and diameter of seminiferous tubules and the seminiferous epithelium (Martin et al. 2010a). In humans and rodents, elevated male BMI is associated with reduced sperm motility (Hammoud et al. 2009), increased incidences of sperm abnormality (Kort et al. 2006a) and sperm DNA fragmentation (Chavarro et al. 2011b), increased sperm reactive oxygen species levels, reduced serum testosterone and increased estradiol concentrations (Tunc et al. 2011). Consumption of a 'Western-style' diet high in sugar, fat and processed food associate with reduced sperm motility in men (Eslamian et al. 2012; Gaskins et al. 2012), while the consumption of energy dense diets in men and rodent models has been associated with poor sperm motility, morphology and DNA integrity (Agbaje et al. 2007), disturbed testis metabolism (Rato et al. 2013) and impaired fertility (Bener et al. 2009). Reduced sperm DNA integrity, as associated with obesity and diabetes, correlates with reduced embryonic development and decreased pregnancy rates (Bakos et al. 2008; Bertolini et al. 2002; Seli et al. 2004).

4.2 Paternal Nutrition and Offspring Development

In men undergoing IVF treatment, obesity is associated with reduced blastocyst development and live birth rates (Bakos et al. 2011a). In mice and rats, the induction of paternal obesity, through the feeding of a high fat diet, increases sperm DNA damage (Bakos et al. 2011b), elevates sperm and testicular oxidative damage (Zhao et al. 2014) and reduces blastocyst development and implantation rates (Mitchell et al. 2011). Paternal dietary-induced obesity has also been shown to induce sub-fertility in both male and female offspring for up to two generations in mice (Fullston et al. 2012, 2013; McPherson et al. 2014). Here, male offspring displayed insulin resistance and hyperleptinaemia, while female offspring became obese with additional insulin resistance. Analysis of F2 generation offspring development revealed increased adiposity and insulin resistance in females, but increased adiposity and glucose intolerance in males. Interestingly, these negative effects on offspring development were prevented through paternal dietary and exercise interventions (McPherson et al. 2013; McPherson et al. 2014; Palmer et al. 2012), indicating sperm-mediated programming may be relatively transient and even reversible. In rats, the feeding of a high fat diet (41% fat) for up to 10 weeks prior to mating affected female offspring pancreatic β-cell function but not males, as well as increased body weight, glucose intolerance and impaired insulin secretion (Ng et al. 2010). Analysis of pancreatic gene transcript levels revealed differential expression of 61 genes involved in ATP binding, intracellular transport, calcium signalling, cell-cycle regulation, apoptosis and MAPK and Wnt signalling pathways. Offspring of male mice overnourished during neonatal life demonstrate glucose intolerance, fasting hyperglycaemia and insulin resistance, mirroring the metabolic disturbance seen in their fathers (Pentinat et al. 2010). Interestingly, F2 generation offspring only displayed mild hyperglycaemia, while neither F1 nor F2 offspring displayed the increased body weight observed in the overfed fathers. As the severity of offspring metabolic programming diminished with subsequent generations, these findings are suggestive of epigenetic mechanisms of gene expression regulation rather than stable, heritable modifications to DNA sequences (Gallou-Kabani and Junien 2005). Recently, Soubry et al. (2013) demonstrated a negative correlation between paternal obesity and the DNA methylation status of the IGF2 differentially methylated region in offspring.

Similar to the impacts of paternal obesity, paternal low protein diet (LPD) programmes the expression of genes involved in offspring hepatic lipid and cholesterol biosynthesis in offspring mice (Carone et al. 2010). Analysis of offspring tissue epigenetic status revealed genome-wide changes in DNA methylation, including the key lipid regulator PPARα. In a rodent model of repeated paternal fasting prior to mating, significantly lowered offspring serum glucose, IGF-1 and corticosterone levels were observed (Anderson et al. 2006). We have shown that offspring from male mice fed LPD display increased weight at birth, reduced male:female offspring ratio, increased adult adiposity, hypotension, glucose intolerance and elevated serum TNF-α levels (Watkins and Sinclair 2014). Recently, the impact of a paternal low folate diet on offspring development has been demonstrated (Lambrot et al. 2013). Sperm from the low folate diet males displayed altered DNA methylation at genes associated with development, apoptosis, autism and schizophrenia. In the placenta, over 300 genes were differentially expressed in response to paternal low folate diet, while increased incidences of offspring craniofacial and musculoskeletal malformations were observed. The impact of parental undernutrition has typically been addressed through the use of supplementary dietary regimens with antioxidant compounds such as folate, zinc and iron (Das et al. 2013). In a recent mouse study (McPherson et al. 2016), the negative impact of a paternal undernutrition diet on sperm quality, testicular oxidative stress, fertility and offspring fat accumulation and dyslipidaemia were reversed through dietary vitamin and antioxidant supplementation. These differing models of paternal dietary manipulation all identify the suboptimally nourished father as a novel source of offspring health programming. In addition, they identify potential intervention strategies which could help alleviate adverse offspring health. Interestingly, some studies demonstrate paternal mediated programming may be reversible through simple dietary supplements of exercise regimens. These observations suggest the process of sperm maturation may be the critically sensitive stage during which paternal programming is established rather than becoming programmed permanently into the spermatogonial stem cells.

4.3 Paternal Toxicant Exposure

In humans, paternal and grand-paternal dietary and smoking behaviours have been shown to influence offspring and grand-offspring phenotype and mortality risk (Pembrey et al. 2006a), while paternal alcoholism is associated with reduced birth weight in offspring (Little and Sing 1987). Exposure of male mice and rats to alcohol impacts on offspring litter size, birth weight, developmental trajectory, mortality, immunity and behavioural characteristics such as spatial awareness, aggression, risk taking and anxiety-like behaviours (Abel 2004; Abel and Bilitzke 1990; Ledig et al. 1998; Meek et al. 2007). Similarly, males exposed to other drugs and environmental toxicants including opiates, ethylene dibromide, lead and cyclophosphamide sire offspring with a range of behavioural impairments, many transmitted to second and third generations via the male lineage (Hales and Robaire 2001).

Similar to nutritional and drug specific effects on paternal programming, endocrine disruptors such as the anti-androgenic compound vinclozolin, have been shown to induce changes in offspring development of exposed fathers.

Offspring from pregnant rat dams exposed to vinclozolin during gestation are at increased risk of tumour formation, kidney disease, immune dysfunction and infertility, transmissible down the male, but not female, lineage for up to four generations (Anway et al. 2005, 2008; Anway and Skinner 2008).

4.4 Molecular and Epigenetic Mechanisms of Paternal Programming

Changes in the normal patterns of sperm DNA (methylation) and histone modifications (methylation, acetylation) or RNA content, delivered at the time of fertilisation, are the prime candidate mechanisms through which altered paternal physiology could impact on offspring development. However, for any paternally established changes in epigenetic status to be faithfully transmitted to adult offspring, any established modifications must escape two major phases of genome-wide epigenetic reprogramming (see Fig. 1). The first phase occurs shortly after fertilisation in

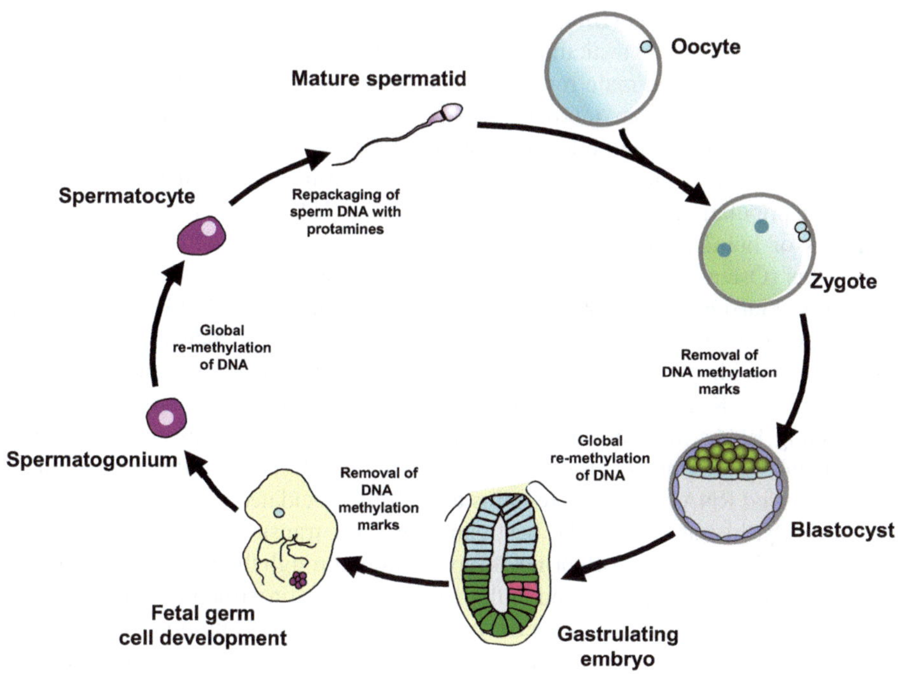

Fig. 1 Diagram representing the major developmental and epigenetic remodelling events during embryo development and sperm maturation. Any paternally derived epigenetic marks must be maintained during two separate phases of genome-wide epigenetic remodelling. Current research is attempting to understand which sperm-derived epigenetic marks are retained during post-fertilisation and how they may affect offspring development

the zygote, when the paternal genome undergoes extensive and active demethylation and remethylation as well as protamine to histone reorganisation (Shi and Wu 2009). The second phase occurs within the migrating primordial germ cells during foetal development. Here, as the primordial germ cells descend down the genital ridge towards the early gonads, genome-wide DNA demethylation occurs in a sex-specific manner (Ly et al. 2015). However, some classes of genes such as imprinted genes and retrotransposable elements have an ability to evade these phases of remodelling, retaining their original epigenetic marks.

Imprinted genes are expressed in a parent-of-origin specific manner, achieved through epigenetic silencing of either the maternal or paternal allele (typically by DNA methylation). Studies in mice have shown that paternal alcohol exposure can modify the DNA methylation status of imprinted genes in sperm (Ouko et al. 2009). Similarly in mice, sperm from male offspring generated through assisted reproductive techniques displayed significant differences in imprinted gene (*Snrpn, H19*) DNA methylation status (Stouder et al. 2009). Exposure of pregnant rat dams to the endocrine disruptor, vinclozolin, results in significant changes in offspring tissue gene expression patterns, for up to four generations (Anway et al. 2005; Stouder and Paoloni-Giacobino 2010). Exposure to drugs such as cocaine and alcohol has been shown to correlate to brain tissue DNA methylation, chromatin conformation and gene expression patterns (Bielawski et al. 2002; Ouko et al. 2009; Pandey et al. 2008) as well as reduced testicular and sperm DNA methyltransferase enzyme transcript levels in exposed male rodents (Bielawski et al. 2002; He et al. 2006; Ouko et al. 2009). Sperm from infertile men have been shown to display significant changes in DNA methylation levels (Aston et al. 2012).

A second transgenerational paternal programming mechanism lies in the extensive population of RNAs (mRNA, siRNA, piRNA) within the mature sperm. Analysis of human sperm has revealed the presence of several thousand coding transcripts (Ostermeier et al. 2002) with their expression profiles linked with male infertility (Garrido et al. 2009; Jodar et al. 2012; Montjean et al. 2012; Platts et al. 2007). Recent studies have employed next-generation sequencing techniques (RNA-Seq) to identify previously uncharacterised and novel coding and small non-coding RNAs (Krawetz et al. 2011; Sendler et al. 2013). The broad spectrum of non-coding RNAs that appear within the mature sperm is suggestive of a role in subsequent development. However, the overall functional significance of these sperm-derived RNA molecules remains to be fully elucidated.

Currently, histones seem the best candidates for transmission of paternal programming effects into the offspring at fertilisation, through their extensive potential for epigenetic modifications and influence on chromatin structure. Infertile men display significant changes in sperm histone populations (Hammoud et al. 2011) and histone methylation (Steilmann et al. 2011; Yap et al. 2011). In mice, haploinsufficiency of sperm protamines lowers sperm counts and induces DNA damage in mice (Cho et al. 2001; Perez-Crespo et al. 2008), while, in humans, altered protamine (P1:P2) ratio associates with reduced fertility rates (Aoki and Carrell 2003; Carrell et al. 2007). Recent analyses of human and mouse sperm active histone (i.e. H3K27me3) localisation revealed significant enrichment at key developmental and pluripotency genes (Brykczynska et al. 2010; Hammoud et al. 2011). While it has yet to be determined whether any of the 2–15% of histones retained within the mammalian sperm contribute directly to zygotic gene expression regulation, studies have revealed that sperm-derived histones are transferred into the oocyte and become incorporated within the zygotic chromatin (van der Heijden et al. 2006; van der Heijden et al. 2008).

In addition to sperm DNA integrity, mRNA populations and histone modifications, sperm are able to influence the developmental programme through the initiation of oocyte calcium oscillations at the point of fertilisation. Upon fertilisation, sperm-derived phospholipase C zeta (PLC-ζ) initiates a series of intracellular Ca^{2+} oscillations that sweep across the egg, lasting for several hours (Swann et al. 2006). Manipulation of the number and amplitude of these Ca^{2+} oscillations has been shown to alter blastocyst cell number (Bos-Mikich et al. 1997) and foetal development (Ozil and Huneau 2001). RNAi knockdown of sperm PLC-ζ in mice has been shown to reduce the number of Ca^{2}+ transients at fertilisation and affect litter size (Knott et al. 2005). Separate to sperm-specific mechanisms of developmental programming, seminal plasma composition (i.e. granulocyte-macrophage colony-stimulating factor) has also been shown to influence embryonic, placental and offspring development (Sjoblom et al. 2005) as well as initiate maternal reproductive tract immunological responses, essential in the establishment and maintenance of pregnancy (Sharkey et al. 2007; Stewart et al. 2009). However, the exact impact of paternal nutrition and physiological status on these additional programming mechanisms, and the long-term offspring cardiovascular and metabolic health risks remain unknown.

5 Conclusions

The concept that parental nutrition during the periconceptional period can have a lasting legacy influencing fertility, offspring health and well-being is now firmly established for a wide range of mammalian species including humans. The study of paternal-mechanisms of transgenerational consequences for offspring health has naturally included an epigenetic perspective. While much of the evidence and support for paternal programming is fundamentally descriptive, advances in sequencing technology are providing scope for analysing epigenetic mechanisms of gene expression regulation and offspring development with increasing detail. However, new models and a greater understanding of the interaction between maternal and paternal genomes during the earliest stages of development are required to explore fully developmental programming mechanisms.

Acknowledgements Dr Watkins is supported by an Aston Research Centre for Healthy Ageing (ARCHA) fellowship and by a Society for Reproduction and Fertility academic scholarship grant.

References

Abel E (2004) Paternal contribution to fetal alcohol syndrome. Addict Biol 9(2):127–133. discussion 135-126. doi:10.1080/13556210410001716980

Abel EL, Bilitzke P (1990) Paternal alcohol exposure: paradoxical effect in mice and rats. Psychopharmacology 100(2):159–164

Agbaje IM, Rogers DA, McVicar CM, McClure N, Atkinson AB, Mallidis C, Lewis SE (2007) Insulin dependant diabetes mellitus: implications for male reproductive function. Hum Reprod 22(7):1871–1877. doi:10.1093/humrep/dem077

Anderson LM, Riffle L, Wilson R, Travlos GS, Lubomirski MS, Alvord WG (2006) Preconceptional fasting of fathers alters serum glucose in offspring of mice. Nutrition 22(3):327–331. doi:10.1016/j.nut.2005.09.006

Anway MD, Skinner MK (2008) Epigenetic programming of the germ line: effects of endocrine disruptors on the development of transgenerational disease. Reprod Biomed Online 16(1):23–25

Anway MD, Cupp AS, Uzumcu M, Skinner MK (2005) Epigenetic transgenerational actions of endocrine disruptors and male fertility. Science 308(5727):1466–1469. doi:10.1126/science.1108190

Anway MD, Rekow SS, Skinner MK (2008) Comparative anti-androgenic actions of vinclozolin and flutamide on transgenerational adult onset disease and spermatogenesis. Reprod Toxicol 26(2):100–106. doi:10.1016/j.reprotox.2008.07.008

Aoki VW, Carrell DT (2003) Human protamines and the developing spermatid: their structure, function, expression and relationship with male infertility. Asian J Androl 5(4):315–324

Aston KI, Punj V, Liu L, Carrell DT (2012) Genome-wide sperm deoxyribonucleic acid methylation is altered in some men with abnormal chromatin packaging or poor in vitro fertilization embryogenesis. Fertil Steril 97(2):285–292. doi:10.1016/j.fertnstert.2011.11.008

Bakos HW, Henshaw RC, Mitchell M, Lane M (2011a) Paternal body mass index is associated with decreased blastocyst development and reduced live birth rates following assisted reproductive technology. Fertil Steril 95(5):1700–1704. doi:10.1016/j.fertnstert.2010.11.044

Bakos HW, Mitchell M, Setchell BP, Lane M (2011b) The effect of paternal diet-induced obesity on sperm function and fertilization in a mouse model. Int J Androl 34(5 Pt 1):402–410. doi:10.1111/j.1365-2605.2010.01092.x

Bakos HW, Thompson JG, Feil D, Lane M (2008) Sperm DNA damage is associated with assisted reproductive technology pregnancy. Int J Androl 31(5):518–526. doi:10.1111/j.1365-2605.2007.00803.x

Barker DJ, Gluckman PD, Godfrey KM, Harding JE, Owens JA, Robinson JS (1993) Fetal nutrition and cardiovascular disease in adult life. Lancet 341(8850):938–941

Bener A, Al-Ansari AA, Zirie M, Al-Hamaq AO (2009) Is male fertility associated with type 2 diabetes mellitus? Int Urol Nephrol 41(4):777–784. doi:10.1007/s11255-009-9565-6

Bertolini M, Mason JB, Beam SW, Carneiro GF, Sween ML, Kominek DJ, Moyer AL, Famula TR, Sainz RD, Anderson GB (2002) Morphology and morphometry of in vivo- and in vitro-produced bovine concepti from early pregnancy to term and association with high birth weights. Theriogenology 58(5):973–994

Bielawski DM, Zaher FM, Svinarich DM, Abel EL (2002) Paternal alcohol exposure affects sperm cytosine methyltransferase messenger RNA levels. Alcohol Clin Exp Res 26(3):347–351

Binder NK, Mitchell M, Gardner DK (2012) Parental diet-induced obesity leads to retarded early mouse embryo development and altered carbohydrate utilisation by the blastocyst. Reprod Fertil Dev 24(6):804–812. doi:10.1071/RD11256

Bos-Mikich A, Whittingham DG, Jones KT (1997) Meiotic and mitotic Ca2+ oscillations affect cell composition in resulting blastocysts. Dev Biol 182(1):172–179. doi:10.1006/dbio.1996.8468

Boucher BJ, Ewen SW, Stowers JM (1994) Betel nut (Areca catechu) consumption and the induction of glucose intolerance in adult CD1 mice and in their F1 and F2 offspring. Diabetologia 37(1):49–55

Brykczynska U, Hisano M, Erkek S, Ramos L, Oakeley EJ, Roloff TC, Beisel C, Schubeler D, Stadler MB, Peters AH (2010) Repressive and active histone methylation mark distinct promoters in human and mouse spermatozoa. Nat Struct Mol Biol 17(6):679–687. doi:10.1038/nsmb.1821

Calle A, Miranda A, Fernandez-Gonzalez R, Pericuesta E, Laguna R, Gutierrez-Adan A (2012) Male mice produced by in vitro culture have reduced fertility and transmit organomegaly and glucose intolerance to their male offspring. Biol Reprod 87(2):34. doi:10.1095/biolreprod.112.100743

Carone BR, Fauquier L, Habib N, Shea JM, Hart CE, Li R, Bock C, Li C, Gu H, Zamore PD, Meissner A, Weng Z, Hofmann HA, Friedman N, Rando OJ (2010) Paternally induced transgenerational environmental reprogramming of metabolic gene expression in mammals. Cell 143(7):1084–1096. doi:10.1016/j.cell.2010.12.008

Carrell DT, Emery BR, Hammoud S (2007) Altered protamine expression and diminished spermatogenesis: what is the link? Hum Reprod Update 13(3):313–327. doi:10.1093/humupd/dml057

Ceelen M, van Weissenbruch MM, Vermeiden JP, van Leeuwen FE, Delemarre-van de Waal HA (2008) Cardiometabolic differences in children born after in vitro fertilization: follow-up study. J Clin Endocrinol Metab 93(5):1682–1688. doi:10.1210/jc.2007-2432

Chavarro JE, Furtado J, Toth TL, Ford J, Keller M, Campos H, Hauser R (2011a) Trans-fatty acid levels in sperm are associated with sperm concentration among men from an infertility clinic. Fertil Steril 95(5):1794–1797. doi:10.1016/j.fertnstert.2010.10.039

Chen TH, Chiu YH, Boucher BJ (2006) Transgenerational effects of betel-quid chewing on the development of the metabolic syndrome in the Keelung community-based integrated screening program. Am J Clin Nutr 83(3):688–692

Cheung YB, Low L, Osmond C, Barker D, Karlberg J (2000) Fetal growth and early postnatal growth are related to blood pressure in adults. Hypertension 36(5):795–800

Cho C, Willis WD, Goulding EH, Jung-Ha H, Choi YC, Hecht NB, Eddy EM (2001) Haploinsufficiency of protamine-1 or −2 causes infertility in mice. Nat Genet 28(1):82–86. doi:10.1038/88313

Dabelea D, Pettitt DJ, Hanson RL, Imperatore G, Bennett PH, Knowler WC (1999) Birth weight, type 2 diabetes, and insulin resistance in pima Indian children and young adults. Diabetes Care 22(6):944–950

Das JK, Salam RA, Kumar R, Bhutta ZA (2013) Micronutrient fortification of food and its impact on woman and child health: a systematic review. Syst Rev 2:67. doi:10.1186/2046-4053-2-67

Dumoulin JC, Land JA, Van Montfoort AP, Nelissen EC, Coonen E, Derhaag JG, Schreurs IL, Dunselman GA, Kester AD, Geraedts JP, Evers JL (2010) Effect of in vitro culture of human embryos on birthweight of newborns. Hum Reprod 25(3):605–612. doi:10.1093/humrep/dep456

Eckert JJ, Porter R, Watkins AJ, Burt E, Brooks S, Leese HJ, Humpherson PG, Cameron IT, Fleming TP (2012) Metabolic induction and early responses of mouse blastocyst developmental programming following maternal low protein diet affecting life-long health. PLoS One 7(12):e52791. doi:10.1371/journal.pone.0052791

Eslamian G, Amirjannati N, Rashidkhani B, Sadeghi MR, Hekmatdoost A (2012) Intake of food groups and idiopathic asthenozoospermia: a case-control study. Hum Reprod 27(11):3328–3336. doi:10.1093/humrep/des311

Fleming TP, Watkins AJ, Sun C, Velazquez MA, Smyth NR, Eckert JJ (2015) Do little embryos make big decisions? How maternal dietary protein restriction can permanently change an embryo. Reprod Fertil Dev 27(4):684–692. doi:10.1071/RD14455

Forsen T, Eriksson JG, Tuomilehto J, Teramo K, Osmond C, Barker DJ (1997) Mother's weight in pregnancy and coronary heart disease in a cohort of Finnish men: follow up study. BMJ 315(7112):837–840

Fullston T, Ohlsson Teague EM, Palmer NO, DeBlasio MJ, Mitchell M, Corbett M, Print CG, Owens JA, Lane M (2013) Paternal obesity initiates metabolic disturbances in two generations of mice with incomplete penetrance to the F2 generation and alters the transcriptional profile of testis and sperm microRNA content. FASEB J : Off Publ Fed Am Societies Exp Biol 27(10):4226–4243. doi:10.1096/fj.12-224048

Fullston T, Palmer NO, Owens JA, Mitchell M, Bakos HW, Lane M (2012) Diet-induced paternal obesity in the absence of diabetes diminishes the reproductive health of two subsequent generations of mice. Hum Reprod 27(5):1391–1400. doi:10.1093/humrep/des030
Gallou-Kabani C, Junien C (2005) Nutritional epigenomics of metabolic syndrome: new perspective against the epidemic. Diabetes 54(7):1899–1906
Garrido N, Martinez-Conejero JA, Jauregui J, Horcajadas JA, Simon C, Remohi J, Meseguer M (2009) Microarray analysis in sperm from fertile and infertile men without basic sperm analysis abnormalities reveals a significantly different transcriptome. Fertil Steril 91(4 Suppl):1307–1310. doi:10.1016/j.fertnstert.2008.01.078
Gaskins AJ, Colaci DS, Mendiola J, Swan SH, Chavarro JE (2012) Dietary patterns and semen quality in young men. Hum Reprod 27(10):2899–2907. doi:10.1093/humrep/des298
Gauthier D, Berbigier P (1982) The influence of nutritional levels and shade structure on testicular growth and hourly variations of plasma-LH and testosterone levels in young creole bulls in a tropical environment. Reprod Nutr Dev 22(5):793–801. doi:10.1051/Rnd:19820606
Grace KS, Sinclair KD (2009) Assisted reproductive technology, epigenetics, and long-term health: a developmental time bomb still ticking. Semin Reprod Med 27(5):409–416. doi:10.1055/s-0029-1237429
Hales BF, Robaire B (2001) Paternal exposure to drugs and environmental chemicals: effects on progeny outcome. J Androl 22(6):927–936
Hammoud AO, Gibson M, Stanford J, White G, Carrell DT, Peterson M (2009) In vitro fertilization availability and utilization in the United States: a study of demographic, social, and economic factors. Fertil Steril 91(5):1630–1635. doi:10.1016/j.fertnstert.2007.10.038
Hammoud SS, Nix DA, Hammoud AO, Gibson M, Cairns BR, Carrell DT (2011) Genome-wide analysis identifies changes in histone retention and epigenetic modifications at developmental and imprinted gene loci in the sperm of infertile men. Hum Reprod 26(9):2558–2569. doi:10.1093/humrep/der192
Hanson MA, Gluckman PD (2014) Early developmental conditioning of later health and disease: physiology or pathophysiology? Physiol Rev 94(4):1027–1076. doi:10.1152/physrev.00029.2013
He F, Lidow IA, Lidow MS (2006) Consequences of paternal cocaine exposure in mice. Neurotoxicol Teratol 28(2):198–209. doi:10.1016/j.ntt.2005.12.003
Heslehurst N, Rankin J, Wilkinson JR, Summerbell CD (2010) A nationally representative study of maternal obesity in England, UK: trends in incidence and demographic inequalities in 619 323 births, 1989-2007. Int J Obes 34(3):420–428. doi:10.1038/ijo.2009.250
Hinkle SN, Sharma AJ, Kim SY, Park S, Dalenius K, Brindley PL, Grummer-Strawn LM (2012) Prepregnancy obesity trends among low-income women, United States, 1999-2008. Matern Child Health J 16(7):1339–1348. doi:10.1007/s10995-011-0898-2
Igosheva N, Abramov AY, Poston L, Eckert JJ, Fleming TP, Duchen MR, McConnell J (2010) Maternal diet-induced obesity alters mitochondrial activity and redox status in mouse oocytes and zygotes. PLoS One 5(4):e10074. doi:10.1371/journal.pone.0010074
Inhorn MC, Patrizio P (2015) Infertility around the globe: new thinking on gender, reproductive technologies and global movements in the 21st century. Hum Reprod Update 21(4):411–426. doi:10.1093/humupd/dmv016
Jodar M, Kalko S, Castillo J, Ballesca JL, Oliva R (2012) Differential RNAs in the sperm cells of asthenozoospermic patients. Hum Reprod 27(5):1431–1438. doi:10.1093/humrep/des021
Jungheim ES, Schoeller EL, Marquard KL, Louden ED, Schaffer JE, Moley KH (2010) Diet-induced obesity model: abnormal oocytes and persistent growth abnormalities in the offspring. Endocrinology 151(8):4039–4046. doi:10.1210/en.2010-0098
Kaati G, Bygren LO, Edvinsson S (2002) Cardiovascular and diabetes mortality determined by nutrition during parents' and grandparents' slow growth period. Eur J Hum Genet 10(11):682–688. doi:10.1038/sj.ejhg.5200859
Kelly T, Yang W, Chen CS, Reynolds K, He J (2008) Global burden of obesity in 2005 and projections to 2030. Int J Obes 32(9):1431–1437. doi:10.1038/ijo.2008.102
Knott JG, Kurokawa M, Fissore RA, Schultz RM, Williams CJ (2005) Transgenic RNA interference reveals role for mouse sperm phospholipase Czeta in triggering Ca2+ oscillations during fertilization. Biol Reprod 72(4):992–996. doi:10.1095/biolreprod.104.036244

Kort HI, Massey JB, Elsner CW, Mitchell-Leef D, Shapiro DB, Witt MA, Roudebush WE (2006a) Impact of body mass index values on sperm quantity and quality. J Androl 27(3):450–452. doi:10.2164/jandrol.05124

Krawetz SA, Kruger A, Lalancette C, Tagett R, Anton E, Draghici S, Diamond MP (2011) A survey of small RNAs in human sperm. Hum Reprod 26(12):3401–3412. doi:10.1093/humrep/der329

Kulie T, Slattengren A, Redmer J, Counts H, Eglash A, Schrager S (2011) Obesity and women's health: an evidence-based review. J Am Board Fam Med 24(1):75–85. doi:10.3122/jabfm.2011.01.100076

Lambrot R, Xu C, Saint-Phar S, Chountalos G, Cohen T, Paquet M, Suderman M, Hallett M, Kimmins S (2013) Low paternal dietary folate alters the mouse sperm epigenome and is associated with negative pregnancy outcomes. Nat Commun 4:2889. doi:10.1038/ncomms3889

Ledig M, Misslin R, Vogel E, Holownia A, Copin JC, Tholey G (1998) Paternal alcohol exposure: developmental and behavioral effects on the offspring of rats. Neuropharmacology 37(1):57–66

Lee HS (2015) Impact of maternal diet on the Epigenome during in utero life and the developmental programming of diseases in childhood and adulthood. Forum Nutr 7(11):9492–9507. doi:10.3390/nu7115467

Lin WY, Chiu TY, Lee LT, Lin CC, Huang CY, Huang KC (2008) Betel nut chewing is associated with increased risk of cardiovascular disease and all-cause mortality in Taiwanese men. Am J Clin Nutr 87(5):1204–1211

Little RE, Sing CF (1987) Father's drinking and infant birth weight: report of an association. Teratology 36(1):59–65. doi:10.1002/tera.1420360109

Ly L, Chan D, Trasler JM (2015) Developmental windows of susceptibility for epigenetic inheritance through the male germline. Semin Cell Dev Biol 43:96–105. doi:10.1016/j.semcdb.2015.07.006

Martin GB, Blache D, Miller DW, Vercoe PE (2010a) Interactions between nutrition and reproduction in the management of the mature male ruminant. Animal 4(7):1214–1226. doi:10.1017/S1751731109991674

McPherson NO, Bakos HW, Owens JA, Setchell BP, Lane M (2013) Improving metabolic health in obese male mice via diet and exercise restores embryo development and fetal growth. PLoS One 8(8):e71459. doi:10.1371/journal.pone.0071459

McPherson NO, Fullston T, Bakos HW, Setchell BP, Lane M (2014) Obese father's metabolic state, adiposity, and reproductive capacity indicate son's reproductive health. Fertil Steril 101(3):865–873. doi:10.1016/j.fertnstert.2013.12.007

McPherson NO, Fullston T, Kang WX, Sandeman LY, Corbett MA, Owens JA, Lane M (2016) Paternal under-nutrition programs metabolic syndrome in offspring which can be reversed by antioxidant/vitamin food fortification in fathers. Sci Rep 6:27010. doi:10.1038/srep27010

Meek LR, Myren K, Sturm J, Burau D (2007) Acute paternal alcohol use affects offspring development and adult behavior. Physiol Behav 91(1):154–160. doi:10.1016/j.physbeh.2007.02.004

Mitchell M, Bakos HW, Lane M (2011) Paternal diet-induced obesity impairs embryo development and implantation in the mouse. Fertil Steril 95(4):1349–1353. doi:10.1016/j.fertnstert.2010.09.038

Mitchell M, Schulz SL, Armstrong DT, Lane M (2009) Metabolic and mitochondrial dysfunction in early mouse embryos following maternal dietary protein intervention. Biol Reprod 80(4):622–630. doi:10.1095/biolreprod.108.072595

Montjean D, De La Grange P, Gentien D, Rapinat A, Belloc S, Cohen-Bacrie P, Menezo Y, Benkhalifa M (2012) Sperm transcriptome profiling in oligozoospermia. J Assist Reprod Genet 29(1):3–10. doi:10.1007/s10815-011-9644-3

Nelissen EC, Van Montfoort AP, Smits LJ, Menheere PP, Evers JL, Coonen E, Derhaag JG, Peeters LL, Coumans AB, Dumoulin JC (2013) IVF culture medium affects human intrauterine growth as early as the second trimester of pregnancy. Hum Reprod 28(8):2067–2074. doi:10.1093/humrep/det131

Ng SF, Lin RC, Laybutt DR, Barres R, Owens JA, Morris MJ (2010) Chronic high-fat diet in fathers programs beta-cell dysfunction in female rat offspring. Nature 467(7318):963–966. doi:10.1038/nature09491

Ombelet W, Cooke I, Dyer S, Serour G, Devroey P (2008) Infertility and the provision of infertility medical services in developing countries. Hum Reprod Update 14(6):605–621. doi:10.1093/humupd/dmn042

Ostermeier GC, Dix DJ, Miller D, Khatri P, Krawetz SA (2002) Spermatozoal RNA profiles of normal fertile men. Lancet 360(9335):772–777. doi:10.1016/S0140-6736(02)09899-9

Ouko LA, Shantikumar K, Knezovich J, Haycock P, Schnugh DJ, Ramsay M (2009) Effect of alcohol consumption on CpG methylation in the differentially methylated regions of H19 and IG-DMR in male gametes: implications for fetal alcohol spectrum disorders. Alcohol Clin Exp Res 33(9):1615–1627. doi:10.1111/j.1530-0277.2009.00993.x

Ozil JP, Huneau D (2001) Activation of rabbit oocytes: the impact of the Ca2+ signal regime on development. Development 128(6):917–928

Palmer NO, Bakos HW, Owens JA, Setchell BP, Lane M (2012) Diet and exercise in an obese mouse fed a high-fat diet improve metabolic health and reverse perturbed sperm function. Am J Phys Endocrinol Metab 302(7):E768–E780. doi:10.1152/ajpendo.00401.2011

Pandey SC, Ugale R, Zhang H, Tang L, Prakash A (2008) Brain chromatin remodeling: a novel mechanism of alcoholism. J Neurosci : Off J Soc Neurosci 28(14):3729–3737. doi:10.1523/JNEUROSCI.5731-07.2008

Parker GV, Thwaites CJ (1972) Effects of Undernutrition on libido and semen quality in adult merino rams. Aust J Agric Res 23(1):109–115. doi:10.1071/Ar9720109

Pembrey ME, Bygren LO, Kaati G, Edvinsson S, Northstone K, Sjostrom M, Golding J (2006a) Sex-specific, male-line transgenerational responses in humans. Eur J Hum Genet : EJHG 14(2):159–166. doi:10.1038/sj.ejhg.5201538

Pentinat T, Ramon-Krauel M, Cebria J, Diaz R, Jimenez-Chillaron JC (2010) Transgenerational inheritance of glucose intolerance in a mouse model of neonatal overnutrition. Endocrinology 151(12):5617–5623. doi:10.1210/en.2010-0684

Perez-Crespo M, Moreira P, Pintado B, Gutierrez-Adan A (2008) Factors from damaged sperm affect its DNA integrity and its ability to promote embryo implantation in mice. J Androl 29(1):47–54. doi:10.2164/jandrol.107.003194

Platts AE, Dix DJ, Chemes HE, Thompson KE, Goodrich R, Rockett JC, Rawe VY, Quintana S, Diamond MP, Strader LF, Krawetz SA (2007) Success and failure in human spermatogenesis as revealed by teratozoospermic RNAs. Hum Mol Genet 16(7):763–773. doi:10.1093/hmg/ddm012

Radlowski EC, Johnson RW (2013) Perinatal iron deficiency and neurocognitive development. Front Hum Neurosci 7:585. doi:10.3389/fnhum.2013.00585

Rato L, Alves MG, Dias TR, Lopes G, Cavaco JE, Socorro S, Oliveira PF (2013) High-energy diets may induce a pre-diabetic state altering testicular glycolytic metabolic profile and male reproductive parameters. Andrology 1(3):495–504. doi:10.1111/j.2047-2927.2013.00071.x

Ravelli AC, van der Meulen JH, Michels RP, Osmond C, Barker DJ, Hales CN, Bleker OP (1998) Glucose tolerance in adults after prenatal exposure to famine. Lancet 351(9097):173–177

Ravelli AC, van Der Meulen JH, Osmond C, Barker DJ, Bleker OP (1999) Obesity at the age of 50 y in men and women exposed to famine prenatally. Am J Clin Nutr 70(5):811–816

Rexhaj E, Paoloni-Giacobino A, Rimoldi SF, Fuster DG, Anderegg M, Somm E, Bouillet E, Allemann Y, Sartori C, Scherrer U (2013) Mice generated by in vitro fertilization exhibit vascular dysfunction and shortened life span. J Clin Invest 123(12):5052–5060. doi:10.1172/JCI68943

Reynolds CM, Gray C, Li M, Segovia SA, Vickers MH (2015) Early life nutrition and energy balance disorders in offspring in later life. Forum Nutr 7(9):8090–8111. doi:10.3390/nu7095384

Rich-Edwards JW, Stampfer MJ, Manson JE, Rosner B, Hankinson SE, Colditz GA, Willett WC, Hennekens CH (1997) Birth weight and risk of cardiovascular disease in a cohort of women followed up since 1976. BMJ 315(7105):396–400

Robinson JJ, Ashworth CJ, Rooke JA, Mitchell LM, McEvoy TG (2006) Nutrition and fertility in ruminant livestock. Anim Feed Sci Technol 126(3–4):259–276. doi:10.1016/j.anifeedsci.2005.08.006

Robker RL, Akison LK, Bennett BD, Thrupp PN, Chura LR, Russell DL, Lane M, Norman RJ (2009) Obese women exhibit differences in ovarian metabolites, hormones, and gene expression compared with moderate-weight women. J Clin Endocrinol Metab 94(5):1533–1540. doi:10.1210/jc.2008-2648

Roseboom TJ, van der Meulen JH, Osmond C, Barker DJ, Ravelli AC, Bleker OP (2000) Plasma lipid profiles in adults after prenatal exposure to the Dutch famine. Am J Clin Nutr 72(5):1101–1106

Roseboom TJ, van der Meulen JH, van Montfrans GA, Ravelli AC, Osmond C, Barker DJ, Bleker OP (2001) Maternal nutrition during gestation and blood pressure in later life. J Hypertens 19(1):29–34

Scherrer U, Rimoldi SF, Rexhaj E, Stuber T, Duplain H, Garcin S, de Marchi SF, Nicod P, Germond M, Allemann Y, Sartori C (2012) Systemic and pulmonary vascular dysfunction in children conceived by assisted reproductive technologies. Circulation 125(15):1890–1896. doi:10.1161/CIRCULATIONAHA.111.071183

Seli E, Gardner DK, Schoolcraft WB, Moffatt O, Sakkas D (2004) Extent of nuclear DNA damage in ejaculated spermatozoa impacts on blastocyst development after in vitro fertilization. Fertil Steril 82(2):378–383. doi:10.1016/j.fertnstert.2003.12.039

Sendler E, Johnson GD, Mao S, Goodrich RJ, Diamond MP, Hauser R, Krawetz SA (2013) Stability, delivery and functions of human sperm RNAs at fertilization. Nucleic Acids Res 41(7):4104–4117. doi:10.1093/nar/gkt132

Sharkey DJ, Macpherson AM, Tremellen KP, Robertson SA (2007) Seminal plasma differentially regulates inflammatory cytokine gene expression in human cervical and vaginal epithelial cells. Mol Hum Reprod 13(7):491–501. doi:10.1093/molehr/gam028

Shi L, Wu J (2009) Epigenetic regulation in mammalian preimplantation embryo development. Reprod Biol Endocrinol : RB&E 7:59. doi:10.1186/1477-7827-7-59

Sinclair KD, Watkins AJ (2013) Parental diet, pregnancy outcomes and offspring health: metabolic determinants in developing oocytes and embryos. Reprod Fertil Dev 26(1):99–114. doi:10.1071/RD13290

Sjoblom C, Roberts CT, Wikland M, Robertson SA (2005) Granulocyte-macrophage colony-stimulating factor alleviates adverse consequences of embryo culture on fetal growth trajectory and placental morphogenesis. Endocrinology 146(5):2142–2153. doi:10.1210/en.2004-1260

Soubry A, Schildkraut JM, Murtha A, Wang F, Huang Z, Bernal A, Kurtzberg J, Jirtle RL, Murphy SK, Hoyo C (2013) Paternal obesity is associated with IGF2 hypomethylation in newborns: results from a newborn epigenetics study (NEST) cohort. BMC Med 11:29. doi:10.1186/1741-7015-11-29

Steilmann C, Paradowska A, Bartkuhn M, Vieweg M, Schuppe HC, Bergmann M, Kliesch S, Weidner W, Steger K (2011) Presence of histone H3 acetylated at lysine 9 in male germ cells and its distribution pattern in the genome of human spermatozoa. Reprod Fertil Dev 23(8):997–1011. doi:10.1071/RD10197

Stewart TM, Liu DY, Garrett C, Brown EH, Baker HW (2009) Recruitment bias in studies of semen and other factors affecting pregnancy rates in fertile men. Hum Reprod 24(10):2401–2408. doi:10.1093/humrep/dep215

Stouder C, Deutsch S, Paoloni-Giacobino A (2009) Superovulation in mice alters the methylation pattern of imprinted genes in the sperm of the offspring. Reprod Toxicol 28(4):536–541. doi:10.1016/j.reprotox.2009.06.009

Stouder C, Paoloni-Giacobino A (2010) Transgenerational effects of the endocrine disruptor vinclozolin on the methylation pattern of imprinted genes in the mouse sperm. Reproduction 139(2):373–379. doi:10.1530/REP-09-0340

Swann K, Saunders CM, Rogers NT, Lai FA (2006) PLCzeta(zeta): a sperm protein that triggers Ca2+ oscillations and egg activation in mammals. Semin Cell Dev Biol 17(2):264–273. doi:10.1016/j.semcdb.2006.03.009

Tunc O, Bakos HW, Tremellen K (2011) Impact of body mass index on seminal oxidative stress. Andrologia 43(2):121–128. doi:10.1111/j.1439-0272.2009.01032.x van der Heijden GW, Derijck AA, Ramos L, Giele M, van der Vlag J, de Boer P (2006) Transmission of modified nucleosomes from the mouse male germline to the zygote and subsequent remodeling of paternal chromatin. Dev Biol 298(2):458–469. doi:10.1016/j.ydbio.2006.06.051

van der Heijden GW, Ramos L, Baart EB, van den Berg IM, Derijck AA, van der Vlag J, Martini E, de Boer P (2008) Sperm-derived histones contribute to zygotic chromatin in humans. BMC Dev Biol 8:34. doi:10.1186/1471-213X-8-34

Vandemark NL, Mauger RE, Fritz GR (1964) Effect of energy intake on reproductive performance of dairy bulls .2. Semen production + replenishment. J Dairy Sci 47(8):898

Wang YC, McPherson K, Marsh T, Gortmaker SL, Brown M (2011) Health and economic burden of the projected obesity trends in the USA and the UK. Lancet 378(9793):815–825. doi:10.1016/S0140-6736(11)60814-3

Watkins AJ, Lucas ES, Wilkins A, Cagampang FR, Fleming TP (2011) Maternal periconceptional and gestational low protein diet affects mouse offspring growth, cardiovascular and adipose phenotype at 1 year of age. PLoS One 6(12):e28745. doi:10.1371/journal.pone.0028745

Watkins AJ, Platt D, Papenbrock T, Wilkins A, Eckert JJ, Kwong WY, Osmond C, Hanson M, Fleming TP (2007) Mouse embryo culture induces changes in postnatal phenotype including raised systolic blood pressure. Proc Natl Acad Sci U S A 104(13):5449–5454. doi:10.1073/pnas.0610317104

Watkins AJ, Sinclair KD (2014) Paternal low protein diet affects adult offspring cardiovascular and metabolic function in mice. Am J Phys Heart Circ Phys 306(10):H1444–H1452. doi:10.1152/ajpheart.00981.2013

Watkins AJ, Ursell E, Panton R, Papenbrock T, Hollis L, Cunningham C, Wilkins A, Perry VH, Sheth B, Kwong WY, Eckert JJ, Wild AE, Hanson MA, Osmond C, Fleming TP (2008) Adaptive responses by mouse early embryos to maternal diet protect fetal growth but predispose to adult onset disease. Biol Reprod 78(2):299–306. doi:10.1095/biolreprod.107.064220

Wrenzycki C, Herrmann D, Lucas-Hahn A, Lemme E, Korsawe K, Niemann H (2004) Gene expression patterns in in vitro-produced and somatic nuclear transfer-derived preimplantation bovine embryos: relationship to the large offspring syndrome? Anim Reprod Sci 82-83:593–603. doi:10.1016/j.anireprosci.2004.05.009

Wu LL, Norman RJ, Robker RL (2011) The impact of obesity on oocytes: evidence for lipotoxicity mechanisms. Reprod Fertil Dev 24(1):29–34. doi:10.1071/RD11904

Yap DB, Walker DC, Prentice LM, McKinney S, Turashvili G, Mooslehner-Allen K, de Algara TR, Fee J, de Tassigny X, Colledge WH, Aparicio S (2011) Mll5 is required for normal spermatogenesis. PLoS One 6(11):e27127. doi:10.1371/journal.pone.0027127

Yeung EH, Druschel C (2013) Cardiometabolic health of children conceived by assisted reproductive technologies. Fertil Steril 99(2):318–326. doi:10.1016/j.fertnstert.2012.12.015

Zhao J, Zhai L, Liu Z, Wu S, Xu L (2014) Leptin level and oxidative stress contribute to obesity-induced low testosterone in murine testicular tissue. Oxidative Med Cell Longev 2014:190945. doi:10.1155/2014/190945

Exploitation of Non-mammalian Model Organisms in Epigenetic Research

William V. Holt

Abstract Model organisms are widely used in research that is ultimately aimed at understanding the causes and consequences of human disease. It may seem counter-intuitive to expect clinically useful information to be obtained from species as diverse as fishes and insects, but because fundamental biological mechanisms share evolutionary origins they transcend species barriers. Epigenetic mechanisms fulfil this expectation admirably as more and more is discovered about the basic operational rules of inheritance, which are much more elaborate than formerly thought. Only a few decades ago, although the complex interplay between genes, inheritance and the environment was recognized, it was difficult to explain. Recent discoveries about the controlling influences of gene silencing through DNA and histone methylation, the roles of so-called non-coding DNA and microRNA, and the way in which these factors respond to environmental conditions have started to shed light on these basic mechanisms. Diverse model species allow epigenetic mechanisms to be studied from different perspectives; for example, some are better suited to studies of sex determination while others may be more convenient for studying the earliest stages of organ development, growth and the influence of nutrition on future wellbeing. The rationale for including this chapter in a book that is focused on uncovering relationships between periconception nutrition in humans is to highlight the opportunities and insights that may be gained by focusing attention on studies in non-mammalian model species.

Keywords Development • Sex determination • Environment • Inheritance • Aquatic organisms • Fishes • Reptiles

W.V. Holt (✉)
Academic Unit of Reproductive and Developmental Medicine, University of Sheffield, Level 4, Jessop Wing, Tree Root Walk, Sheffield S10 2SF, UK
e-mail: Bill.Holt@sheffield.ac.uk

A. Fazeli, W.V. Holt (eds.), *Periconception in Physiology and Medicine*, Advances in Experimental Medicine and Biology 1014,
DOI 10.1007/978-3-319-62414-3_9

1 Introduction

Model organisms have been studied extensively in biology and have provided a wealth of information on basic physiological and developmental processes. In principle, because species from multiple genera share common, albeit often very ancient, evolutionary origins with humans, studies of these organisms can produce valuable information that is directly useful in understanding human diseases. The choice of model species is usually related to the topic under investigation, the facilities available to researchers, rapidity of growth and reproduction and indeed the expertise of the researchers. Model organisms can also range across orders of scale, from bacteria, yeasts, nematodes and protozoa to large farm animals; all have their place in biological research. In fact, a systematic study of animal model organisms used in studies of development and covering the years 1965, 1975, 1985, 1995 and 2005 (Davies 2007) showed that model species came from 287 genera. There was, however, a strong bias towards the mouse and fruit fly, which together accounted for 40% of the 4615 papers examined.

Recent developments in high throughput genomics and proteomics have, however, called into question the value of working with animal models when research into human diseases can frequently be conducted directly upon human subjects (Fields and Johnston 2005). Indeed, the increasing emphasis on replacing, reducing and refining the use of research animals (known as the 3Rs; Graham and Prescott 2015) is positively encouraging scientists to explore methods that avoid the use of whole animals. Developmental biology is, however, such a conserved process that model organisms will undoubtedly continue to provide a wealth of information that is relevant to humans. Advances in our understanding of reproductive sciences and developmental biology have recently converged in ways that blur the boundaries between traditionally recognized biological disciplines. It is also apparent that understanding the extent to which developmental processes are affected by various aspects of an organism's environment is increasingly important. Organismal responses to the environment appear to share common mechanisms across species and genera, and model organisms are able to provide basic information that illuminate the entire scope of the subject.

Organisms of all types tend to adapt and evolve in order to make the best of their situation and the adaptive responses often involve subtle, or even major, phenotypic changes that can occur surprisingly quickly, even from one generation to the next (Gluckman et al. 2009; Bateson et al. 2014). Such short-term processes, considered to be forms of developmental plasticity, do not necessitate changes in DNA sequence, but instead rely on the controlled modulation of gene expression pathways using epigenetic mechanisms (Burggren and Crews 2014) such as DNA methylation and histone modification (Schaefer and Nadeau 2015) that were not even recognized until relatively recently (Daxinger and Whitelaw 2010). Understanding developmental plasticity as a conserved process is therefore highly relevant to unravelling mechanisms that affect human health, and the objective of this review is to highlight the way in which a few key non-mammalian model species are well suited when it comes to exploring such complexities.

2 Environmental Effects on Reproduction and Development

Developmental programmes and embryonic growth involve some degree of forecasting the state of the future environment so that embryo development can be matched against expectations (Nichelmann 2004). In fact, this relationship is at the heart of reproductive seasonality, where species have developed physiological strategies that typically synchronize breeding and parturition with climatic conditions for the optimal survival of their offspring (Bronson 2009). If the forecast turns out to be inaccurate because the environment changes and defies expectations, the mismatch can have unfortunate consequences for adult health. If the expectation does not match the reality, the outcome often results in disease; such relationships are now well recognized in humans, owing mainly to a series of studies linking early life nutrition to the onset of diseases in adulthood (see, e.g. Barker et al. 2010; Kwong et al. 2000). The realization that there are systematic relationships between pre- and/or periconception nutritional conditions and chronic disease conditions in adults has stimulated a great deal of research in humans and other species. In fact, the subject is now widely known as "DoHad" or the "Developmental Origins of Health and Disease", and there is even a specific journal devoted to the topic. The scope of the DoHad concept is ever widening, as exemplified by a recent article (Ferraro et al. 2016) in that journal (*Journal of Developmental Origins of Health and Disease*) which links air pollution, caesarean section and domestic violence during pregnancy into the DoHad paradigm. Transgenerational effects on body condition in males caused by father's and grandfather's smoking habits have also been described (Pembrey et al. 2006), as well as, in rats, the transgenerational impacts of experimentally administered agricultural chemicals (Anway et al. 2006a, 2006b). Our current understanding of DoHad therefore encompasses the effects of anthropogenic environmental chemicals on humans and animals (Haugen et al. 2015), and a discussion of model organisms would be incomplete if these aspects were omitted.

The word "environment" covers a multiplicity of factors in which organisms survive; it is apparent that the developmental plasticity exhibited by various species may be responses to changes in nutrition, food availability, population density, the prevalence of infectious diseases, presence of competitors and many other stresses. Uncovering epigenetic mechanisms across a wide range of species is therefore highly instructive for understanding details about the ways in which evolution progresses. Knowledge of transcriptomics was not available to Darwin or Lamarck when they formulated their concepts of evolution and selection, and it is therefore very valuable to keep evolutionary biology in mind as novel epigenetic mechanisms come to light (Van Soom et al. 2014).

3 Fishes as Model Organisms

Nutritional support is provided to mammalian embryos via their complex interactions with the placenta with the result that, with the exception of monotremes, live offspring are born; this process is known as viviparity. Analogous but diverse embryonic support mechanisms leading to the birth of live offspring are believed to have evolved about 150 times in different vertebrate clades (Blackburn 2015), including fishes and amphibians. There are more than 25,000 species of fish, of which about 500 are viviparous (Wourms et al. 1988). Their reproductive adaptations are highly variable and do not always involve the provisioning of embryos via placenta-like structures. However, in the context of DoHad and the investigation of periconception nutrition, it is worth mentioning two groups of fishes, the Poeciliidae and the Syngnathidae, as potential research models.

The Poeciliidae are a family of around 200 small tropical fish species (Pires et al. 2010), some of which have evolved internal fertilization and placental modes of embryo support. These are already used extensively as valuable model species in reproductive studies and in studies on the way in which their development is affected by environmental pollution. It is of interest here that insulin-like growth factor 2 (IGF2), which is involved in the regulation of mammalian embryonic growth, has been found to be expressed in two species of Poeciliidae, *Heterandria formosa* and *Poeciliopsis prolifica* (Lawton et al. 2005; O'Neill et al. 2007). Placental morphology in the Poeciliidae has been studied in some detail (Kwan et al. 2015; Schindler 2015), and the impact upon ovarian function and fecundity of feeding diets that vary in protein content (Dahlgren 1980) was investigated over 30 years ago. However, because these fishes are considered to be well suited for laboratory research (Petrescu-Mag et al. 2014), they offer an excellent opportunity for novel investigations.

A series of experiments aimed at investigating relationships between reproductive outcomes and the quality of food supplied experimentally to female sailfin mollies (*Poecilia latipinna*) prior to ovulation (Trexler 1997), showed that if large females received low quality diet, they responded to their circumstance by providing a greater degree of placental support to their embryos than females receiving high quality diet. One member of the Poeciliidae (*Phalloptychus januarius*) that shows a particularly high level of matrotrophy (i.e. post-fertilization embryo provisioning) has also been used as a model species to investigate relationships between maternal diet, embryo survival and growth (Pollux and Reznick 2011). In this particular experiment, there was no evidence for embryo abortion under low food conditions, but developing offspring were small at birth. The value of these experiments lies not only in the insights they provide about the Poeciliidae themselves but also the insights into the underlying fundamental biology that transcends the phylogenetic divisions. As noted elsewhere in this book, the immediate and future developmental potential of embryos is significantly determined by maternal nutritional status, both before and during pregnancy, and being able to explore these interactions using a small and rapidly reproducing viviparous species provides an efficient means of investigating the principles involved.

The Poeciliidae are unusual in that they have also evolved the phenomenon of superfetation, whereby females can carry several independently fertilized broods of embryos at the same time (for review, see Pollux et al. 2009). Although this ability has been investigated for its evolutionary and ecological benefits, few if any developmental studies have explored the complex nutritional experiments that may be possible.

Seahorses and pipefishes are another group of teleostian candidate model organisms, the *Syngnathidae*, that hold considerable promise for exploring the effects of periconceptional nutrition on embryonic growth and survival. Seahorse and pipefish embryos, like those of most fishes, possess a yolk sac as a source of nutrients. However, unlike most fishes, these embryos develop within a brood pouch, which is a specialized body cavity only found in males (Schmid and Senn 2002; Stölting and Wilson 2007). Recent radioisotope studies have demonstrated that male pipefishes contribute towards embryonic nutrition via the brood pouch (Ripley and Foran 2009; Kvarnemo et al. 2011; Kornienko 2001); the embryos become embedded within well-vascularized (Ripley et al. 2010) depressions of the interior lining of the brood pouch (Carcupino et al. 2002), and the fertilized embryos develop to term within a closed environment (Fig. 1). Thus, the brood pouch of the male is becoming more widely regarded as functionally equivalent to a mammalian uterus. Indeed, recent genomic studies have shown that the seahorse/pipefish pouch expresses genes more commonly associated with the mammalian uterus and oviduct (Whittington et al. 2015). Despite the intensive degree of husbandry needed for successful seahorse and pipefish breeding in the laboratory, these species hold considerable promise as model organisms suitable for the study of the periconception environment in relation to offspring health and survival.

As seahorses are widely exhibited in zoos and aquaria, there is a substantial body of research on the merits of diets and feeding regimes (Wong and Benzie 2003;

Fig. 1 Photograph of a male seahorse (*Hippocampus hippocampus*) in its natural environment off the coast of Gran Canaria, Spain. The prominent brood pouch indicates that this individual is pregnant. Scale bar = 1 cm (Photograph courtesy of Elodie Turpin)

Woods 2000b, a, 2002, 2003); however, relatively little attention has been paid to the experimental study of pre- and periconception diet in relation to embryonic growth and survival. One study (Otero-Ferrer et al. 2014) in which male and female *Hippocampus reidi* were fed separately with different qualities of diet ("wild caught" or "commercially available") showed, however, that preconception dietary effects could be distinguished on the basis of gender. When females that had only received commercial diet prior to mating were mated with males that had received the natural, wild-caught, diet, the resultant embryos showed poor survival after birth, coupled with distortion of relative head and snout size. When, conversely, males had been provided only with commercial diet prior to mating with females that had received wild-caught diet, the resultant offspring were abnormally large and also showed poor survival. The main difference between diets concerned the content of polyunsaturated fatty acids, with the commercial feed being somewhat, but not enormously, less enriched. In principle these results showed that seahorses could potentially be used as model organisms, where, because of their biology, it is possible to distinguish between dietary effects induced via oocyte quality (thereby mediated solely by females) and those affected by the quality of pouch (placental) support, which are mediated via males. This contrasts with the difficulty of separating oocyte-specific and placenta-specific effects of periconception diets in mammals and other viviparous vertebrates, because any diet fed to females would normally affect both ovarian and uterine function. The possibility that male-mediated dietary effects are influenced by sperm quality as well as pouch function is still something to respect; however, ingenious experimental designs could overcome this problem and allow researchers more flexibility than can be achieved using mammals.

Environmental influences on fish development have long been a topic of intense interest to researchers (reviewed by (Jonsson and Jonsson 2014)). Multiple aspects of the life history of fishes, including growth and adult body size, sex ratio, egg size, lifespan and tendency to migrate, are affected by early influences and although the mechanisms responsible are poorly understood, it is becoming ever more apparent that epigenetics is a central theme. As an example, using sea bass (*Dicentrarchus labrax)* as a model species, Navarro-Martin et al. (2011) demonstrated that sex ratio skewing in this species, where higher temperatures during a critical period in early development induce higher proportions of male offspring, are associated with increased DNA methylation of the aromatase promoter. The temperature-dependent sex ratio skew was therefore attributed to the prevention of aromatase gene expression via an epigenetic mechanism. Recent studies have demonstrated that epigenetic mechanisms are also involved in the temperature-dependency of embryonic growth in fishes via modulation of DNA methylation and gene expression. For example, thermal stress in Atlantic cod embryos resulted in stage-specific changes in the expression profiles of 5-methyltransferase enzymes (Skjaerven et al. 2014), while temperature changes altered the expression of microRNAs during embryonic development in Senegalese sole (*Solea senegalensis*) (Campos et al. 2014); these changes were associated with variations in growth rate, muscle development and lipid metabolism.

The zebrafish (*Danio rerio*) is widely used as a model organism for developmental and molecular studies, partly because it is relatively easy to manage and breed in the laboratory, and partly because of the ease with which the early stages of development can be viewed microscopically. Despite the huge number of published studies that have focused on zebrafish development, the mechanisms that drive sex determination remain stubbornly unclear. Elements of both genetic and temperature-dependent sex determination have been identified (Lopez-Olmeda and Sanchez-Vazquez 2011; Abozaid et al. 2012; Liew et al. 2012; Takatsu et al. 2013; Liew and Orban 2014; Shen and Wang 2014), with a bias towards the development of males at higher temperatures. Development of gender itself is unexpectedly complex, whereby the initial formation of immature ovaries in all individuals is followed by masculinization and testicular development in the case of males. This level of complexity is almost certainly driven epigenetically, but at present the mechanisms are unknown. Sex determination in marine species such as the bluehead wrasse (Fig. 2) (*Thalassoma bifasciatum*) is also a complex issue, where individuals initially develop as females but under specific circumstances can turn into males after reaching adulthood; the socially controlled mechanism is known as protogyny. To make the situation more complicated there are two male phenotypes, distinguished by both body size and colour markings. Recent transcriptomic studies comparing male and female gonads and brains have shown that sex differences are correlated with major changes in gonadal gene expression (Liu et al. 2015). The authors believed that the sex determining pathways in this species were not notably

Fig. 2 Photograph of a male *bluehead* wrasse (*Thalassoma bifasciatum*), a species found on coral reefs in the Caribbean sea. These fishes reach a maximum length of 25 cm, with a maximum reported age of 3 years, but individuals can reach sexual maturity when they are less than 4 cm in length. Males develop the striking *blue* head once they are sexually mature, but a small proportion of adult females also develop this coloration when they undergo a sex change induced by their social environment (Photograph courtesy of Jeff Whitlock; http://www.theonlinezoo.com)

different from those found in other vertebrates, which presumably means that signals emanating from the social environment are sufficient to cause the sexual plasticity.

4 Amphibians and Epigenetics

The amphibians are a group of species that includes the frogs, toads, newts, salamanders and caecilians. One of the most remarkable characteristics of the urodele amphibians, i.e. the newts and salamanders, is the ability to regenerate parts of the body such as the limbs, tail, jaw, retina and even brain, via a process known as epimorphic regeneration (Grigoryan 2016). Regenerative abilities are not completely absent in other species (Godwin 2014), even deer regenerate their antlers annually prior to the breeding season, but some amphibians have evolved this ability to an exceptional degree. One of the most fascinating but puzzling aspects of tissue regeneration in certain amphibians is their ability to replicate an entire limb, with all of the appropriate cell and tissue types in their correct places. Such an ability implies that cells possess an intrinsic "positional memory" which directs cell division and differentiation. While such a hypothesis has been around for many years, it is only now that the likely role of epigenetic mechanisms is being identified. For example, studies of *Xenopus tropicalis* (Hayashi et al. 2015) have identified histone acetylation and deacetylation as important epigenetic control mechanisms in the determination of positional memory. One of the longest established animal models in biology is the axolotl (Fig. 3) (*Ambystoma mexicanum*), which is a species of salamander from the lakes around Mexico City (Reiss et al. 2015). All amphibians pass through a larval stage, where they use gills for respiration, before metamorphosing into adults, but the axolotl is exceptional in that it rarely goes through this process and spends its life in the larval stage even after reaching sexual maturity. Axolotls also possess exceptional abilities when it comes to limb regeneration and can regenerate an entire limb in 100–200 days, depending on species. Positional memory is clearly essential for the regeneration of entire limbs, given that nerves, cartilage,

Fig. 3 Photograph of an Axolotl (*Ambystoma mexicanum),* a salamander from the lakes around Mexico City. Individuals of this species typically reach about 30 cm in length and remain in the larval stage throughout their adult life (Photograph courtesy of Jeff Whitlock; http://www.theonlinezoo.com)

muscle and blood vessels have to be correctly assembled. A recent study by Kragl et al. (2009) showed that the cells at the site of injury rapidly organize themselves into a new site for growth. The various cell types retain their original identity and are not pluripotent, and grow synchronously in order to form the new limb or other organ. Such observations show that the axolotl is a perfect model species for a wide range of epigenetic studies. Nevertheless, there are still relatively few publications dealing with epigenetic processes and tissue regeneration in axolotls. Data on the expression of microRNA during tail regeneration in the axolotl with directly valuable implications for vertebrate development were published recently by Gearhart et al. (2015). These authors identified 4564 microRNA families known to be widely conserved among vertebrates, as well as 59,811 reads of putative novel microRNAs. The authors considered that their data supported the hypothesis that microRNAs play key roles in managing the precise spatial and temporal patterns of gene expression during tissue repair and development. Other authors (Aguilar and Gardiner 2015) have also confirmed that tissue repair requires subtle modification of the DNA methylation status, which changes quickly after injury, following changes in the expression of DNA methyltransferases. Amphibians exhibit many other usual aspects of developmental plasticity that are almost certainly epigenetically controlled; one recent report has shown that the capacity of *Xenopus laevis* to survive prolonged periods without food is brought about by an ability to reduce gene transcription in the intestine and thus to depress metabolic rate (Tamaoki et al. 2016).

Xenopus tropicalis, as well as its larger cousin *Xenopus laevis*, are widely used as a model species in biological research and toxicology, and now rival the rat, mouse and zebrafish in their usage. Breeding facilities such as the European Xenopus Resource Centre (EXRC) in Portsmouth, UK, supply animals to other researchers in biomedical research, and even provide them with genetic resources such as frozen spermatozoa and unfrozen oocytes.

5 Aquatic Organisms as Model Species in Reproductive Toxicology

Some species of the fish group Poeciliidae have been used as bio-indicators of pollution in aquatic environments, especially with respect to the detection and effects of endocrine disrupting chemicals (EDCs). A study of Eastern mosquitofish (*Gambusia holbrooki*) in the St Johns River, Florida (Bortone and Cody 1999), found that females showed significant elongation of the anal fin and the gonopodium, an anal fin that is modified into an intromittent organ in male Poeciliidae. A similar study in China that focused on another mosquitofish (*Gambusia affinis*) (Hou et al. 2011) found evidence of masculinizing effects on the anal fin but also detected increased testis mass. Interestingly, these study sites both received effluents from a local paper mill. Paper mill effluents have since been implicated as causing female masculinization in other sites (Deaton and Cureton 2011). Multiple other effects of endocrine disrupting chemicals have been detected using Poeciliidae, including changes in female mating behaviour, altered offspring sex ratios, diminished body size in masculinized females and lower fecundity (Tian et al. 2012; Sharbidre and Sopanrao Patode 2012; Vigario and Saboia-Morais 2014; da Silva et al. 2014; Shenoy 2014). One study has even reported changes in risk taking behaviour (Heintz et al. 2015) induced in male and female *Poecilia reticulata* individuals by the estrogenic compound, 17-ethinylestradiol and the androgenic compound, 17-trenbolone.

The Poeciliidae are not, of course, the only group of aquatic organisms that exhibit sensitivity to endocrine disrupting chemicals. Many environmental studies of wild fish have demonstrated male feminization and female masculinization mediated via the presence of anthropogenic compounds, including heavy metals, plasticisers, flame retardants, polychlorinated biphenyls, that typically show some ability to bind, albeit weakly, to hormone receptors (for reviews, see Kumar and Holt 2014; Bhandari et al. 2015; Wilkinson et al. 2016).

Apart from fishes, it is also noteworthy that many aquatic invertebrates are used, or can be used, as model organisms because they also respond to EDCs in the environment (European Environment Agency 2012; Kumar and Holt 2014). Molluscs appear to possess oestrogen-like receptors, but these are apparently not activated by the vertebrate oestrogen, estradiol, or by other known vertebrate EDCs. Nevertheless, mud snails responded to 12.5% and 25% sewage by increased embryo production, while higher sewage concentration (50%) reduced it (Jobling et al. 2004). Later studies (Sternberg et al. 2008) suggested that mud snails do, in fact, respond to androgens and oestrogens, but through an alternative signalling route, possibly the retinoid system. Similar studies on sewage exposure have detected increased vitellogenin-like proteins in males, feminized sex ratios and low gonadosomatic indices (Chesman and Langston 2006; Gomes et al. 2009). Potent androgen receptor agonists and aromatase inhibitors, as well as the marine anti-fouling paint component tributyltin (TBT), induce “imposex” in female gastropod molluscs at concentrations as low as parts per billion. This is where the penis “imposes” on the normal female anatomy, blocking the oviduct and inducing sterility (Pascoal et al. 2013). As knowledge of epigenetics in marine organisms is growing fast, the use of next-generation sequencing combined with studies of epigenetic mechanisms in marine invertebrates has recently been proposed as a source of rapid and powerful biomonitoring tools for tracking the quality of the marine environment (Suarez-Ulloa et al. 2015).

6 Sex Determination and Epigenetics

Environmentally controlled sex determination occurs in both vertebrates (Valenzuela and Lance 2004) and invertebrates (Adams et al. 1987), being affected by photoperiod, ambient temperature, availability of resources, populations density and even, in the nematode *Mermis nigrescens*, the size of the grasshopper that hosts it (Craig and Webster 1982). Temperature-dependent sex determination (TSD) in reptiles is well documented and has been studied for many years, but until recently it has been difficult to explain mechanistically. Recent advances in genomics have begun to implicate epigenetic mechanisms, however, and therefore although it would be impractical to regard crocodiles, alligators and turtles as model species, it is worth mentioning them here. A recent study of temperature effects (Yatsu et al. 2016) on the gonadal transcriptome in American alligator (*Alligator mississippiensis*) embryos incubated at male and female producing temperatures of 33.5 and 30 °C, respectively, revealed sex-dependent expression of 230 genes, 25 of which were involved in transcriptional regulation. The authors also reported the sexually dimorphic expression of non-coding RNA. In keeping with observations in mammals that environmental factors can impact subsequent generations, similarly long-lasting effects have been induced experimentally in an Australian reptile, the jacky dragon *Amphibolurus muricatus*, whereby the offspring sex ratio is affected by the incubation temperature experienced by the father several years earlier (Warner et al. 2013). In this study the eggs were incubated at female producing temperatures (23 and 33 °C) and an intermediate temperature (27 °C) that produces both sexes. Applying an aromatase inhibitor to 50% of the eggs in each treatment resulted in the production of males, even at female producing temperatures; the other 50% were used as controls. The essential outcome of this study was that the males produced forcibly by the combination of low temperature and aromatase went on to produce predominantly male offspring when they bred about 3 years later. Interestingly, the offspring sex ratio produced by fathers forcibly bred at the higher temperature was skewed in the opposite direction, i.e. towards the production of females.

Zebrafish (*Danio rerio*), although not viviparous, are widely used for laboratory research into the influence of endocrine disrupting chemicals (EDCs) on development. A simple search for "zebrafish" + "endocrine*" + "disrupt*" on "Web of Science" detected more than 1200 publications describing various aspects of such research. A major advantage of the zebrafish over other research models stems from the wealth of background information about development and the role of epigenetics already available for this species (see, e.g. Saxena et al. 2016; Liu et al. 2016). Although there is still relatively little known about the way in which EDCs impact on epigenetic processes during zebrafish development, this is a growing field. In a recent review, Kamstra et al. (2015) discussed DNA methylation at length, reporting that zebrafish possess eight highly conserved DNA methyltransferase genes. As DNA methylation and demethylation occurs during zebrafish development, and these processes are affected by a range of toxicants, it is clear that the impacts of environmental contaminants must be mediated via epigenetic mechanisms, and that endocrine effects are likely to be downstream of these initial impacts.

Water fleas, such as *Daphnia*, which is a genus of freshwater crustacean, occur naturally within the zooplankton assemblages found in bodies of freshwater. They have long been a favourite model species for biologists interested in developmental plasticity because of their ability to modify their life history traits in response to environmental changes. Many species respond to the presence of predators by developing head and tail spines, and can accelerate their developmental rate, switch between sexual and asexual reproductive modes and change the sex ratios of their offspring as appropriate. *Daphnia* have therefore become an important group of model species for studies in aquatic toxicology (Shaw et al. 2008). However, *Daphnia* are also a valuable organism for research in epigenetics because they parthenogenetically produce individuals, which are essentially clones of the parent, and display phenotypic variation that depends on environmental conditions (for review, see (Bonasio 2015)). Given these attributes, it is not surprising that the *Daphnia* genome has now been sequenced.

Sexual differentiation in *Daphnia* is controlled by the environment (LeBlanc and Medlock 2015) and environmental cues such as season (and therefore ambient temperature), availability of food resources, population density and loss of habitat are known to be involved. The mechanisms of sex determination are known to involve neuropeptides, ecdysone and methyl farnesoate (Toyota et al. 2015; La et al. 2014) and a genetic component also seems likely. However, despite the extensive body of work the sex determination mechanism remains stubbornly unclear. Nevertheless, recent studies have documented the existence of both histone modifications and DNA methylation in *Daphnia*, and therefore it seems likely that this group of species will become an excellent model for future epigenetic studies (Robichaud et al. 2012; Harris et al. 2012).

7 Epigenetics and Social Insects

Social insects provide an unusual opportunity to examine the putative involvement of epigenetics in development because despite fertilized eggs often being produced by a single female, the queen, in a given colony, the variation in eventual phenotypes is astonishing. Not only do the offspring differentiate as males or females, usually depending on the season, but the adults develop highly varied behaviours and phenotypes (Yan et al. 2015). As an example, the workers in a honey bee (*Apis mellifera*) colony are all female and their role is in foraging for food and caring for the developing larvae. However, each year a group of males is produced whose sole duty is to emerge from the colony and mate with a queen, thereby providing the queen with spermatozoa that are then stored in the female reproductive tract for a prolonged period and used for fertilization. While the queen is capable of mating and producing fertilized eggs, the female workers are sterile and their ovaries do not develop. This form of social organization, with a caste system, has contributed to the global success of social insects, leading some authors to draw parallels between insect and human societies in terms of the division of labour (Middleton and Latty 2016). Over the last decade there has been considerable interest in the possibility that the various phenotypes within the insect caste systems are controlled through epigenetic mechanisms (Maleszka 2008; Yan et al. 2015; Yan et al. 2014). The early development of Honey bees (*Apis mellifera*) seems to be determined very early in life, starting from the time when the larvae receive different quality diets from workers (worker jelly) or from the queen (royal jelly). Royal jelly contains factors that have a direct impact on DNA methylation status; for example, knockdown of an

Apis mellifera DNA methyltransferase (*Dnmt3*) by RNA interference results in the development of a queen phenotype from larvae destined to become workers.

Investigators studying the composition of worker jelly have shown that it is highly enriched in microRNA species compared to royal jelly (Guo et al. 2013) and, moreover, that the nature of the microRNA changes on a daily basis during development. Such observations, coupled with data showing that royal jelly is also rich in 10-hydroxy-2-decenoic acid (Spannhoff et al. 2011; Wang et al. 2014) (sometimes known as queen bee acid), which comprises up to 5% of royal jelly and can inhibit histone deacetylation, is suggestive of an epigenetic basis for the development of caste-specific phenotypes. Although these observations relate specifically to honey bees, the hypothesis that morphology and behaviour in social insect societies are controlled via epigenetic mechanisms, and especially through DNA methylation (Standage et al. 2016), is receiving ever more support (Sumner et al. 2006; Patalano et al. 2012; Maleszka 2008; Weiner et al. 2013).

8 Conclusions

Embryonic development always begins with a single cell that contains a large, but arguably finite, number of "genes", and the estimated number of genes in a human embryo is believed to be somewhere between 20 and 25,000. Cell division ensues shortly after fertilization and eventually a multitude of different tissue and cell types are produced by growth and differentiation. Even the cells within a single organ are demonstrably different from each other in terms of their morphology, function and transcriptome; the gonads provide a good example, with morphologically distinct germ cells, specialized somatic cells with and without endocrine functionality and a diversity of biochemical functions and responses. How can this impressively enormous amount of variability be squeezed out of a finite number of genes? In the early days of genomic research the definition of a gene was fairly straightforward; each protein was believed to be encoded by a single DNA sequence and gene expression was controlled via regulatory feedback mechanisms (Beckwith 2011). The number of genes needed for producing phenotypic variability was therefore believed to be correspondingly high, possibly more than 100,000 for a mammal. However, more recently "the gene" has become a much more complex concept as it now incorporates the many permutations by which parts of DNA sequences can be expressed. The timing of gene expression is also crucially important in development and day to day living; gene expression is capable of dynamic control, being modulated by extrinsic and intrinsic factors such as day length and circadian rhythm. It is therefore apparent from the foregoing discussion that many, or indeed all, species have evolved regulatory mechanisms to control the way in which DNA sequences are transcribed and translated. These mechanisms, which, because they are often modulated by small RNA molecules, are themselves encoded within the organism's own genome, thus suggesting that the evolution of species involves selective forces that act upon the epigenetic regulatory mechanisms. Such musings lead towards the hypothesis that epigenetic mechanisms must be continuously evolving in tandem with environmental influences, always striving to predict and adapt to future conditions in which a new generation will develop, and promoting the selection and survival of well adapted species. While this is not an original hypothesis, it is one that has so far not received much attention (Smith et al. 2016); however, it will without doubt provide a rich and novel topic for future investigations. It will probably also provide novel insights into ways that the environment controls many aspects of evolution and integrate more molecular thinking into future ecological and evolutionary research.

References

Abozaid H, Wessels S, Hoerstgen-Schwark G (2012) Elevated temperature applied during gonadal transformation leads to male bias in zebrafish (*Danio rerio*). Sexual Development 6(4):201–209. doi:10.1159/000336297

Adams J, Greenwood P, Naylor C (1987) Evolutionary aspects of environmental sex determination. Int J Invertebr Reprod Dev 11(2):123–135. doi:10.1080/01688170.1987.10510273

Aguilar C, Gardiner DM (2015) DNA methylation dynamics regulate the formation of a regenerative wound epithelium during axolotl limb regeneration. PLoS One 10(8):e0134791. doi:10.1371/journal.pone.0134791

Anway MD, Leathers C, Skinner MK (2006a) Endocrine disruptor vinclozolin induced epigenetic transgenerational adult-onset disease. Endocrinology 147(12):5515–5523

Anway MD, Memon MA, Uzumcu M, Skinner MK (2006b) Transgenerational effect of the endocrine disruptor vinclozolin on male spermatogenesis. J Androl 27(6):868–879

Barker DJ, Gelow J, Thornburg K, Osmond C, Kajantie E, Eriksson JG (2010) The early origins of chronic heart failure: impaired placental growth and initiation of insulin resistance in childhood. Eur J Heart Fail 12(8):819–825. hfq069 [pii]10.1093/eurjhf/hfq069

Bateson P, Gluckman P, Hanson M (2014) The biology of developmental plasticity and the Predictive Adaptive Response hypothesis. J Physiol-London 592(11):2357–2368. doi:10.1113/jphysiol.2014.271460

Beckwith J (2011) The Operon as paradigm: normal science and the beginning of biological complexity. J Mol Biol 409(1):7–13. http://dx.doi.org/10.1016/j.jmb.2011.02.027

Bhandari RK, Deem SL, Holliday DK, Jandegian CM, Kassotis CD, Nagel SC, Tillitt DE, Saal FSV, Rosenfeld CS (2015) Effects of the environmental estrogenic contaminants bisphenol A and 17 alpha-ethinyl estradiol on sexual development and adult behaviors in aquatic wildlife species. Gen Comp Endocrinol 214:195–219. doi:10.1016/j.ygcen.2014.09.014

Blackburn DG (2015) Evolution of vertebrate viviparity and specializations for fetal nutrition: a quantitative and qualitative analysis. J Morphol 276(8):961–990. doi:10.1002/jmor.20272

Bonasio R (2015) The expanding epigenetic landscape of non-model organisms. J Exp Biol 218(1):114–122. doi:10.1242/jeb.110809

Bortone SA, Cody RP (1999) Morphological masculinization in poeciliid females from a paper mill effluent receiving tributary of the St. Johns River, Florida. USA Bull Environ Contam Toxicol 63(2):150–156

Bronson FH (2009) Climate change and seasonal reproduction in mammals. Philos Trans R Soc Lond Ser B Biol Sci 364(1534):3331–3340. doi:10.1098/rstb.2009.0140

Burggren WW, Crews D (2014) Epigenetics in comparative biology: why we should pay attention. Integr Comp Biol 54(1):7–20. doi:10.1093/icb/icu013

Campos C, Sundaram AYM, Valente LMP, Conceicao LEC, Engrola S, Fernandes JMO (2014) Thermal plasticity of the miRNA transcriptome during Senegalese sole development. BMC Genomics 15:1525. doi:10.1186/1471-2164-15-525

Carcupino M, Baldacci A, Mazzini M, Franzoi P (2002) Functional significance of the male brood pouch in the reproductive strategies of pipefishes and seahorses: a morphological and ultrastructural comparative study on three anatomically different pouches. J Fish Biol 61:1465–1480

Chesman BS, Langston WJ (2006) Intersex in the clam Scrobicularia plana: a sign of endocrine disruption in estuaries? Biol Lett 2(3):420–422. doi:10.1098/rsbl.2006.0482

Craig SM, Webster JM (1982) Influence of host nutrients on the parasitic development of Mermis nigrescens (Mermithidae). J Nematol 14(3):398

da Silva ES, Moreno Abril SI, Zanette J, Bianchini A (2014) Salinity-dependent copper accumulation in the guppy *Poecilia vivipara* is associated with CTR1 and ATP7B transcriptional regulation. Aquat Toxicol 152:300–307. doi:10.1016/j.aquatox.2014.04.024

Dahlgren BT (1980) The effects of 3 different dietary-protein levels on the fecundity in the guppy, *poecilia reticulata* (PETERS). J Fish Biol 16(1):83–97. doi:10.1111/j.1095-8649.1980.tb03688.x

Davies JA (2007) Developmental biologists' choice of subjects approximates to a power law, with no evidence for the existence of a special group of 'model organisms'. BMC Dev Biol 7(1):1–7. doi:10.1186/1471-213x-7-40

Daxinger L, Whitelaw E (2010) Transgenerational epigenetic inheritance: more questions than answers. Genome Res 20(12):1623–1628. doi:10.1101/gr.106138.110

Deaton R, Cureton JC II (2011) Female masculinization and reproductive life history in the western mosquitofish (*Gambusia affinis*). Environ Biol Fish 92(4):551–558. doi:10.1007/s10641-011-9878-z

European Environment Agency (ed) (2012) The impact of endocrine disrupters on wildlife, people and their environments. The Weybridge +15 (1996–2011) report. 2. EEA Technical report, Copenhagen

Ferraro AA, Fernandes MTB, Vieira SE (2016) New challenges beyond nutrition: C-section, air pollution and domestic violence. J Dev Orig Health Dis 7(03):253–256. doi:10.1017/S204017441500149X

Fields S, Johnston M (2005) Cell biology. Whither model organism research? Science 307(5717): 1885–1886. doi:10.1126/science.1108872

Gearhart MD, Erickson JR, Walsh A, Echeverri K (2015) Identification of conserved and novel micrornas during tail regeneration in the Mexican axolotl. Int J Mol Sci 16(9):22046–22061. doi:10.3390/ijms160922046

Gluckman PD, Hanson MA, Bateson P, Beedle AS, Law CM, Bhutta ZA, Anokhin KV, Bougneres P, Chandak GR, Dasgupta P, Smith GD, Ellison PT, Forrester TE, Gilbert SF, Jablonka E, Kaplan H, Prentice AM, Simpson SJ, Uauy R, West-Eberhard MJ (2009) Towards a new developmental synthesis: adaptive developmental plasticity and human disease. Lancet 373(9675):1654–1657. doi:10.1016/S0140-6736(09)60234-8

Godwin J (2014) The promise of perfect adult tissue repair and regeneration in mammals: learning from regenerative amphibians and fish. BioEssays 36(9):861–871. doi:10.1002/bies.201300144

Gomes T, Gonzalez-Rey M, Bebianno MJ (2009) Incidence of intersex in male clams *Scrobicularia plana* in the Guadiana estuary (Portugal). Ecotoxicology 18(8):1104–1109. doi:10.1007/s10646-009-0359-5

Graham ML, Prescott MJ (2015) The multifactorial role of the 3Rs in shifting the harm-benefit analysis in animal models of disease. Eur J Pharmacol 759:19–29. http://dx.doi.org/10.1016/j.ejphar.2015.03.040

Grigoryan EN (2016) High regenerative ability of tailed amphibians (Urodela) as a result of the expression of juvenile traits by mature animals. Russ J Dev Biol 47(2):83–92. doi:10.1134/s1062360416020041

Guo XQ, Su SK, Skogerboe G, Dai SAJ, Li WF, Li ZG, Liu F, Ni RF, Guo Y, Chen SL, Zhang SW, Chen RS (2013) Recipe for a busy bee: microRNAs in honey bee caste determination. PLoS One 8(12):e81661. doi:10.1371/journal.pone.0081661

Harris KD, Bartlett NJ, Lloyd VK (2012) Daphnia as an emerging epigenetic model organism. Genet Res Int 2012:147892. doi:10.1155/2012/147892

Haugen AC, Schug TT, Collman G, Heindel JJ (2015) Evolution of DOHaD: the impact of environmental health sciences. J Dev Orig Health Dis 6(2):55–64. doi:10.1017/S2040174414000580

Hayashi S, Kawaguchi A, Uchiyama I, Kawasumi-Kita A, Kobayashi T, Nishide H, Tsutsumi R, Tsuru K, Inoue T, Ogino H, Agata K, Tamura K, Yokoyama H (2015) Epigenetic modification maintains intrinsic limb-cell identity in Xenopus limb bud regeneration. Dev Biol 406(2):271–282. doi:10.1016/j.ydbio.2015.08.013

Heintz MM, Brander SM, White JW (2015) Endocrine Disrupting Compounds Alter Risk-Taking Behavior in Guppies (*Poecilia reticulata*). Ethology 121(5):480–491. doi:10.1111/eth.12362

Hou L, Xie Y, Ying G, Fang Z (2011) Developmental and reproductive characteristics of western mosquitofish (*Gambusia affinis*) exposed to paper mill effluent in the Dengcun river, Sihui, South China. Aquat Toxicol 103(3–4):140–149. doi:10.1016/j.aquatox.2011.02.018

Jobling S, Casey D, Rogers-Gray T, Oehlmann J, Schulte-Oehlmann U, Pawlowski S, Baunbeck T, Turner AP, Tyler CR (2004) Comparative responses of molluscs and fish to environmental estrogens and an estrogenic effluent. Aquat Toxicol 66(2):207–222

Jonsson B, Jonsson N (2014) Early environment influences later performance in fishes. J Fish Biol 85(2):151–188. doi:10.1111/jfb.12432

Kamstra JH, Aleström P, Kooter JM, Legler J (2015) Zebrafish as a model to study the role of DNA methylation in environmental toxicology. Environ Sci Pollut Res 22(21):16262–16276. doi:10.1007/s11356-014-3466-7

Kornienko E (2001) Reproduction and development in some genera of pipefish and seahorses of the family Syngnathidae. Russ J Mar Biol 27(1):S15–S26

Kragl M, Knapp D, Nacu E, Khattak S, Maden M, Epperlein HH, Tanaka EM (2009) Cells keep a memory of their tissue origin during axolotl limb regeneration. Nature 460(7251):60–U69. doi:10.1038/nature08152

Kumar E, Holt WV (2014) Impacts of endocrine disrupting chemicals on reproduction in wildlife. Adv Exp Med Biol 753:55–70. doi:10.1007/978-1-4939-0820-2_4

Kvarnemo C, Mobley KB, Partridge C, Jones AG, Ahnesjo I (2011) Evidence of paternal nutrient provisioning to embryos in broad-nosed pipefish *Syngnathus typhle*. J Fish Biol 78(6):1725–1737. doi:10.1111/j.1095-8649.2011.02989.x

Kwan L, Fris M, Rodd FH, Rowe L, Tuhela L, Panhuis TM (2015) An examination of the variation in maternal placentae across the genus Poeciliopsis (*Poeciliidae*). J Morphol 276(6):707–720. doi:10.1002/jmor.20381

Kwong WY, Wild AE, Roberts P, Willis AC, Fleming TP (2000) Maternal undernutrition during the preimplantation period of rat development causes blastocyst abnormalities and programming of postnatal hypertension. Development 127(19):4195–4202

La GH, Choi JY, Chang KH, Jang MH, Joo GJ, Kim HW (2014) Mating behavior of daphnia: impacts of predation risk, food quantity, and reproductive phase of females. PLoS One 9(8):e104545. doi:10.1371/journal.pone.0104545

Lawton BR, Sevigny L, Obergfell C, Reznick D, O'Neill RJ, O'Neill MJ (2005) Allelic expression of IGF2 in live-bearing, matrotrophic fishes. development genes and. Evolution 215(4):207–212. doi:10.1007/s00427-004-0463-8

LeBlanc GA, Medlock EK (2015) Males on demand: the environmental-neuro-endocrine control of male sex determination in daphnids. FEBS J 282(21):4080–4093. doi:10.1111/febs.13393

Liew WC, Bartfai R, Lim Z, Sreenivasan R, Siegfried KR, Orban L (2012) Polygenic sex determination system in zebrafish. PLoS One 7(4):e34397. doi:10.1371/journal.pone.0034397

Liew WC, Orban L (2014) Zebrafish sex: a complicated affair. Brief Funct Genomics 13(2):172–187. doi:10.1093/bfgp/elt041

Liu H, Lamm MS, Rutherford K, Black MA, Godwin JR, Gemmell NJ (2015) Large-scale transcriptome sequencing reveals novel expression patterns for key sex-related genes in a sex-changing fish. Biol Sex Differ 6:26. doi:10.1186/s13293-015-0044-8

Liu Y, Zhang Y, Tao S, Guan Y, Zhang T, Wang Z (2016) Global DNA methylation in gonads of adult zebrafish *Danio rerio* under bisphenol A exposure. Ecotoxicol Environ Saf 130:124–132

Lopez-Olmeda JF, Sanchez-Vazquez FJ (2011) Thermal biology of zebrafish (*Danio rerio*). J Therm Biol 36(2):91–104. doi:10.1016/j.jtherbio.2010.12.005

Maleszka R (2008) Epigenetic integration of environmental and genomic signals in honey bees. Epigenetics 3(4):188–192

Middleton EJT, Latty T (2016) Resilience in social insect infrastructure systems. J R Soc Interface 13(116). doi:10.1098/rsif.2015.1022

Navarro-Martin L, Vinas J, Ribas L, Diaz N, Gutierrez A, Di Croce L, Piferrer F (2011) DNA methylation of the gonadal aromatase (cyp19a) promoter Is involved in temperature-dependent sex ratio shifts in the European sea bass. PLoS Genet 7(12):e1002447. doi:10.1371/journal.pgen.1002447

Nichelmann M (2004) Perinatal epigenetic temperature adaptation in avian species: comparison of turkey and Muscovy duck. J Therm Biol 29(7):613–619

O'Neill MJ, Lawton BR, Mateos M, Carone DM, Ferreri GC, Hrbek T, Meredith RW, Reznick DN, O'Neill RJ (2007) Ancient and continuing Darwinian selection on insulin-like growth factor II in placental fishes. Proc Natl Acad Sci U S A 104(30):12404–12409. doi:10.1073/pnas.0705048104

Otero-Ferrer F, Izquierdo M, Fazeli A, Holt WV (2014) Embryonic developmental plasticity in the long-snouted seahorse (*Hippocampus reidi*, Ginsburg 1933) in relation to parental preconception diet. Reprod Fertil Dev. doi:10.1071/RD14169

Pascoal S, Carvalho G, Vasieva O, Hughes R, Cossins A, Fang Y, Ashelford K, Olohan L, Barroso C, Mendo S, Creer S (2013) Transcriptomics and *in vivo* tests reveal novel mechanisms underlying endocrine disruption in an ecological sentinel, *Nucella lapillus*. Mol Ecol 22(6):1589–1608. doi:10.1111/mec.12137

Patalano S, Hore TA, Reik W, Sumner S (2012) Shifting behaviour: epigenetic reprogramming in eusocial insects. Curr Opin Cell Biol 24(3):367–373. doi:10.1016/j.ceb.2012.02.005

Pembrey ME, Bygren LO, Kaati G, Edvinsson S, Northstone K, Sjostrom M, Golding J (2006) Sex-specific, male-line transgenerational responses in humans. Eur J Hum Genet 14(2):159–166

Petrescu-Mag IV, Oroian IG, Tomescu R, Esanu VO (2014) Laboratory fish strains and species of fish suitable for water quality research. ProEnvironment 7(17):26–29

Pires MN, Arendt J, Reznick DN (2010) The evolution of placentas and superfetation in the fish genus Poecilia (Cyprinodontiformes: Poeciliidae: subgenera Micropoecilia and Acanthophacelus). Biol J Linn Soc 99(4):784–796. doi:10.1111/j.1095-8312.2010.01391.x

Pollux BJA, Pires MN, Banet AI, Reznick DN (2009) Evolution of placentas in the fish family poeciliidae: an empirical study of macroevolution. Annu Rev Ecol Evol Syst 40:271–289. doi:10.1146/annurev.ecolsys.110308.120209

Pollux BJA, Reznick DN (2011) Matrotrophy limits a female's ability to adaptively adjust offspring size and fecundity in fluctuating environments. Funct Ecol 25(4):747–756. doi:10.1111/j.1365-2435.2011.01831.x

Reiss C, Olsson L, Hossfeld U (2015) The history of the oldest self-sustaining laboratory animal: 150 years of axolotl research. J Exp Zoo B-Mol Dev Evol 324(5):393–404. doi:10.1002/jez.b.22617

Ripley JL, Foran CM (2009) Direct evidence for embryonic uptake of paternally-derived nutrients in two pipefishes (*Syngnathidae*: Syngnathus spp.) J Comp Physiol B 179(3):325–333. doi:10.1007/s00360-008-0316-2

Ripley JL, Williams PS, Foran CM (2010) Morphological and quantitative changes in paternal brood-pouch vasculature during embryonic development in two Syngnathus pipefishes. J Fish Biol 77(1):67–79. doi:10.1111/j.1095-8649.2010.02659.x

Robichaud NF, Sassine J, Beaton MJ, Lloyd VK (2012) The epigenetic repertoire of daphnia magna includes modified histones. Genet Res Int 2012:174860. doi:10.1155/2012/174860

Saxena S, Purushothaman S, Meghah V, Bhatti B, Poruri A, Meena Lakshmi MG, Sarath Babu N, Narasimha Murthy CL, Mandal KK, Kumar A, Idris MM (2016) Role of annexin gene and its regulation during zebrafish caudal fin regeneration. Wound repair and regeneration 24(3):551–559. doi:10.1111/wrr.12429

Schaefer S, Nadeau JH (2015) The genetics of epigenetic inheritance: Modes, Molecules, and Mechanisms. Q Rev Biol 90(4):381–415

Schindler JF (2015) Structure and function of placental exchange surfaces in goodeid fishes (Teleostei: Atheriniformes). J Morphol 276(8):91–1003. doi:10.1002/jmor.20427
Schmid MS, Senn DG (2002) Seahorses - Masters of adaptation. Vie Milieu 52(4):201–207
Sharbidre AA, Sopanrao Patode P (2012) Behavioural changes and acetylcholinesterase activity in guppy fish (*Poecilia reticulata*) exposed to chlorpyrifos. Nat Environ Pollut Technol 11(3):487–492
Shaw JR, Pfrender ME, Eads BD, Klaper R, Callaghan A, Sibly RM, Colson I, Jansen B, Gilbert D, Colbourne JK (2008) Daphnia as an emerging model for toxicological genomics. In: Hogstrand C, Kille P (eds) Comparative toxicogenomics, advances in experimental biology. 2 165–219 doi:10.1016/s1872-2423(08)00005-7
Shen Z-G, Wang H-P (2014) Molecular players involved in temperature-dependent sex determination and sex differentiation in Teleost fish. Genet Sel Evol 46:26. doi:10.1186/1297-9686-46-26
Shenoy K (2014) Prenatal exposure to low doses of atrazine affects mating behaviors in male guppies. Horm Behav 66(2):439–448. doi:10.1016/j.yhbeh.2014.07.002
Skjaerven KH, Hamre K, Penglase S, Finn RN, Olsvik PA (2014) Thermal stress alters expression of genes involved in one carbon and DNA methylation pathways in Atlantic cod embryos. Comp Biochem Physiol A-Mol & Integr Physiol 173:17–27. doi:10.1016/j.cbpa.2014.03.003
Smith TA, Martin MD, Nguyen M, Mendelson TC (2016) Epigenetic divergence as a potential first step in darter speciation. Mol Ecol 25(8):1883–1894. doi:10.1111/mec.13561
Spannhoff A, Kim YK, Raynal NJM, Gharibyan V, Su MB, Zhou YY, Li J, Castellano S, Sbardella G, Issa JPJ, Bedford MT (2011) Histone deacetylase inhibitor activity in royal jelly might facilitate caste switching in bees. EMBO Rep 12(3):238–243. doi:10.1038/embor.2011.9
Standage DS, Berens AJ, Glastad KM, Severin AJ, Brendel VP, Toth AL (2016) Epigenetics studies in ecology and evolution Genome, transcriptome and methylome sequencing of a primitively eusocial wasp reveal a greatly reduced DNA methylation system in a social insect. Mol Ecol 25(8):1769–1784. doi:10.1111/mec.13578
Sternberg RM, Hotchkiss AK, LeBlanc GA (2008) The contribution of steroidal androgens and estrogens to reproductive maturation of the eastern mud snail Ilyanassa obsoleta. Gen Comp Endocrinol 156(1):15–26. http://dx.doi.org/10.1016/j.ygcen.2007.12.002
Stölting KN, Wilson AB (2007) Male pregnancy in seahorses and pipefish: beyond the mammalian model. BioEssays 29(9):884–896. doi:10.1002/bies.20626
Suarez-Ulloa V, Gonzalez-Romero R, Eirin-Lopez JM (2015) Environmental epigenetics: A promising venue for developing next-generation pollution biomonitoring tools in marine invertebrates. Mar Pollut Bull 98(1–2):5–13. doi:10.1016/j.marpolbul.2015.06.020
Sumner S, Pereboom JJM, Jordan WC (2006) Differential gene expression and phenotypic plasticity in behavioural castes of the primitively eusocial wasp, *Polistes canadensis*. Proc R Soc B Biol Sci 273(1582):19–26. doi:10.1098/rspb.2005.3291
Takatsu K, Miyaoku K, Roy SR, Murono Y, Sago T, Itagaki H, Nakamura M, Tokumoto T (2013) Induction of female-to-male sex change in adult zebrafish by aromatase inhibitor treatment. Sci Rep 3(3400). doi:10.1038/srep03400
Tamaoki K, Okada R, Ishihara A, Shiojiri N, Mochizuki K, Goda T, Yamauchi K (2016) Morphological, biochemical, transcriptional and epigenetic responses to fasting and refeeding in intestine of *Xenopus laevis*. Cell and Bioscience 6(2). doi:10.1186/s13578-016-0067-9
Tian H, Li Y, Wang W, Wu P, Ru S (2012) Exposure to monocrotophos pesticide during sexual development causes the feminization/demasculinization of the reproductive traits and a reduction in the reproductive success of male guppies (*Poecilia reticulata*). Toxicol Appl Pharmacol 263(2):163–170. doi:10.1016/j.taap.2012.06.006
Toyota K, Miyakawa H, Hiruta C, Furuta K, Ogino Y, Shinoda T, Tatarazako N, Miyagawa S, Shaw JR, Iguchi T (2015) Methyl farnesoate synthesis is necessary for the environmental sex determination in the water flea Daphnia pulex. J Insect Physiol 80:22–30. doi:10.1016/j.jinsphys.2015.02.002
Trexler JC (1997) Resource availability and plasticity in offspring provisioning: embryo nourishment in sailfin mollies. Ecology 78(5):1370–1381

Valenzuela N, Lance V (2004) Temperature-dependent sex determination in vertebrates. Smithsonian Books Washington, DC

Van Soom A, Peelman L, Holt WV, Fazeli A (2014) An introduction to epigenetics as the link between genotype and environment: a personal view. Reprod Domest Anim 49(Suppl 3):2–10. doi:10.1111/rda.12341

Vigario AF, Saboia-Morais SMT (2014) Effects of the 2,4-D herbicide on gills epithelia and liver of the fish *Poecilia vivipara*. Pesqui Vet Bras 34(6):523–528

Wang WX, Tian LQ, Huang Q, Wu XB, Zeng ZJ (2014) Effects of 10-Hydroxy-2-decenoic acid on the development of honey bee (*Apis mellifera*) larvae. J Apic Res 53(1):171–176. doi:10.3896/ibra.1.53.1.19

Warner DA, Uller T, Shine R (2013) Transgenerational sex determination: the embryonic environment experienced by a male affects offspring sex ratio. Scientific Reports 3:4–4

Weiner SA, Galbraith DA, Adams DC, Valenzuela N, Noll FB, Grozinger CM, Toth AL (2013) A survey of DNA methylation across social insect species, life stages, and castes reveals abundant and caste-associated methylation in a primitively social wasp. Naturwissenschaften 100(8):795–799. doi:10.1007/s00114-013-1064-z

Whittington CM, Griffith OW, Qi W, Thompson MB, Wilson AB (2015) Seahorse brood pouch transcriptome reveals common genes associated with vertebrate pregnancy. Mol Biol Evol 32(12):3114–3131. doi:10.1093/molbev/msv177

Wilkinson JL, Hooda PS, Barker J, Barton S, Swinden J (2016) Ecotoxic pharmaceuticals, personal care products, and other emerging contaminants: a review of environmental, receptor-mediated, developmental, and epigenetic toxicity with discussion of proposed toxicity to humans. Crit Rev Environ Sci Technol 46(4):336–381. doi:10.1080/10643389.2015.1096876

Wong JM, Benzie JAH (2003) The effects of temperature, Artemia enrichment, stocking density and light on the growth of juvenile seahorses, *Hippocampus whitei* (Bleeker, 1855), from Australia. Aquaculture 228(1–4):107–121

Woods CMC (2000a) Improving initial survival in cultured seahorses, *Hippocampus abdominalis* Leeson, 1827 (Teleostei : Syngnathidae). Aquaculture 190(3–4):377–388

Woods CMC (2000b) Preliminary observations on breeding and rearing the seahorse *Hippocampus abdominalis* (Teleostei : Syngnathidae) in captivity. N Z J Mar Freshw Res 34(3):475–485

Woods CMC (2002) Natural diet of the seahorse *Hippocampus abdominalis*. N Z J Mar Freshw Res 36(3):655–660

Woods CMC (2003) Growth and survival of juvenile seahorse *Hippocampus abdominalis* reared on live, frozen and artificial foods. Aquaculture 220(1–4):287–298

Wourms JP, Grove BD, Lombardi J (1988) 1 The maternal-embryonic relationship in viviparous fishes. In: Hoar WS, Randall DJ (eds) Fish Physiology 11 Part B 1-134.doi:http://dx.doi.org/10.1016/S1546-5098(08)60213-7

Yan H, Bonasio R, Simola DF, Liebig J, Berger SL, Reinberg D (2015) DNA methylation in social Insects: how epigenetics can control behavior and longevity. In: Berenbaum MR (ed) Annual Review of Entomology, 60:435–452. doi:10.1146/annurev-ento-010814-020803

Yan H, Simola DF, Bonasio R, Liebig J, Berger SL, Reinberg D (2014) Eusocial insects as emerging models for behavioural epigenetics. Nat Rev Genet 15(10):677–688. doi:10.1038/nrg3787

Yatsu R, Miyagawa S, Kohno S, Parrott BB, Yamaguchi K, Ogino Y, Miyakawa H, Lowers RH, Shigenobu S, Guillette LJ, Iguchi T (2016) RNA-seq analysis of the gonadal transcriptome during *Alligator mississippiensis* temperature-dependent sex determination and differentiation. BMC Genomics 17(1):1–13. doi:10.1186/s12864-016-2396-9

Index

A. Fazeli, W.V. Holt (eds.), *Periconception in Physiology and Medicine*,
Advances in Experimental Medicine and Biology 1014,
DOI 10.1007/978-3-319-62414-3

MIX
Papier aus verantwortungsvollen Quellen
Paper from responsible sources
FSC® C105338

If you have any concerns about our products,
you can contact us on
ProductSafety@springernature.com

In case Publisher is established outside the EU,
the EU authorized representative is:
Springer Nature Customer Service Center GmbH
Europaplatz 3, 69115 Heidelberg, Germany

Printed by Libri Plureos GmbH
in Hamburg, Germany